HOW TO
MASTER
SCIENCE LABS

HOW TO MASTER SCIENCE LABS

DIANE A. WALLACE
AND PHILIP L. HERSHEY

AN EXPERIMENTAL
SCIENCE SERIES BOOK

FRANKLIN WATTS
NEW YORK / LONDON
TORONTO / SYDNEY / 1987

All photographs courtesy of Salvatore Tocci
except Carolina Biological Supply Company:
p. 85 (bottom left and right);
Dennis Milon: p. 89; NASA: p. 114.

Diagrams by Vantage Art

Library of Congress Cataloging-in-Publication Data

Wallace, Diane.
How to master science labs.

(An Experimental science series book)
Includes index.
Summary: Explains methods and techniques used
in lab experiments, covering such topics as heating, measuring, and collecting substances, doing
dissections, using lab equipment, and gathering
data. Includes instructions for actual experiments.
1. Science—Experiments—Juvenile literature.
2. Physical laboratories—Juvenile literature.
3. Physical measurements—Juvenile literature.
[1. Science—Experiments. 2. Experiments]
I. Hershey, Philip L. II. Title. III. Series
Q164.W24 1987   507'.24   86-24733
ISBN 0-531-15124-7 (paper ed.)
ISBN 0-531-10323-4 (lib. bdg.)

Copyright © 1987 by Diane A. Wallace and Philip L. Hershey
All rights reserved
Printed in the United States of America
6 5 4 3

CONTENTS

Preface
11

Chapter One
Believe It or Not . . .
Science Labs
Are Important!
13

Chapter Two
The Recipe
for Success
19

Chapter Three
How to Succeed
on Lab Day
27

Chapter Four
The Tools and Techniques
of the Trade
45

Chapter Five
R2D2 Has Nothing
on You
93

Chapter Six
You're Ready to Do
a Science Project
107

Glossary
123

Index
125

LAB TECHNIQUE QUICK REFERENCE GUIDE

TECHNIQUES FOR MIXING AND DISPENSING AND FOR WORKING WITH GLASS 46
How to Work with Glassware 46
How to Mix Solutions 48
How to Pour a Liquid
   from a Reagent Bottle 52
How to Use a Mortar and Pestle 54
How to Cut and Bend Glass Tubing
   and Solid Glass Rods 55
How to Insert Glass
   into a Rubber Stopper 58

TECHNIQUES FOR MEASURING 60
How to Use a Balance 60
How to Use a Compass 62
How to Use a Protractor 63
How to Use a Metric Ruler 64
How to Use Graduated Cylinders 66
How to Use a Thermometer 68

TECHNIQUES FOR HEATING SUBSTANCES 70
How to Use a Laboratory Burner 70
How to Heat Substances in a Test Tube 72

TECHNIQUES FOR COLLECTING AND SEPARATING SUBSTANCES 74
How to Use an Evaporating Dish 74
How to Use a Funnel and Filter Paper 76
How to Collect a Gas 78
How to Identify Oxygen, Hydrogen, and Carbon Dioxide 80

TECHNIQUES FOR OBSERVING 82
How to Dissect Plants and Animals 82
How to Use a Microscope 84
How to Use a Telescope 88
How to Work with a Petri Dish 90

HOW TO MASTER SCIENCE LABS

PREFACE

Amy and Justin are laboratory partners at William Penn High School. Today in chemistry lab they are assigned the task of identifying an unknown chemical. Amy enters the lab with great enthusiasm. She gathers her lab material, sets up the equipment, prepares her data sheets, and begins to experiment. She is businesslike in her approach and is enjoying herself, lost in the challenge of science.

Justin stands to the side and watches. He is reluctant to join in the activities because he suffers from a disease common to science students: "lab fright." This disease attacks students as the lab period arrives. Students who lack the proper lab skills are its prime victims. The symptoms are disorganization, confusion, and ever-increasing frustration. They ask, "What is really going on in this lab?"

This book is designed to prevent or cure lab fright. It will help you to become a more successful lab student who enjoys the challenge of science.

1
BELIEVE IT OR NOT...
SCIENCE LABS
ARE IMPORTANT!

Science isn't just your fourth-period class. Nor is it a spectator sport. Science affects every aspect of your life. Science makes possible the passage from ideas to actions. Quiet, enduring experimentation in the laboratory has led to astounding applications that have created entirely new industries and have provided for better living. As Louis Pasteur said in the early 1800s, "Science is the soul of the prosperity of nations and the living source of all progress."

Scientific research in the laboratory is responsible for improved tools, clothing, and shelter, and for the improved health, safety, and nutrition of a good proportion of the human race. Research in the medical field provides but one example. The vaccination has eliminated the threat of smallpox, polio, and other illnesses that once caused fear, pain, and even death for millions of people. Discoveries in preventive medicine, nutrition, and physical fitness have helped to keep the human body healthy.

All modern conveniences are the result of scientific discoveries. Each time a switch is flicked to illuminate a room or a golf cart rides over the fairways, thanks can be given to Michael Faraday and Joseph Henry. Their scientific experimentation led to the development of the electric power system and the electric motor.

The VCR, microcomputers, laser disks, solar panels, disease-resistant plants, test tube babies, artificial human organs, genetically engineered human hormones, energy-independent homes, nuclear reactors, space stations, space travel, moon bases, orbiting industrial plants—the list of applications of discoveries made in the science laboratory goes on and on.

WHY SCIENTISTS EXPERIMENT

Scientific work is done for many reasons, but usually not for a grade or credit. Greater rewards, such as the thrill of discovery, often encourage the scientist to persevere. Most scientific experimentation has developed from the trial-and-error methods of early humans to the more precise procedures of today. While early scientific approaches were unorganized, today most scientists follow a well-established scientific method to direct experimentation.

Scientists experiment to seek new knowledge, solve problems, test hypotheses (educated guesses), or confirm theories. These quests into the unknown are often motivated by curiosity. Modern society stimulates discovery by creating competition. All research, however, stems from the common goal of improving the quality of human life.

Scientists experiment for various reasons. Gregor Mendel, an Austrian monk, developed a hypothesis about how traits were passed from parent to offspring. To prove this idea he experimented with the garden pea. Seven years of experimenting enabled Mendel to show mathematically that his hypothesis was correct. Mendel could in many cases almost predict the outcome of a particular cross before it took place. Today scientists still experiment to prove their hypotheses.

Sometimes scientists experiment to find solutions to existing problems. Louis Pasteur made an observation that milkmaids who contracted cowpox were somehow immune to getting a more serious disease, smallpox. Through experimentation, Pasteur developed the idea of immunization. He developed a smallpox vaccine. Pasteur demonstrated the application of the scientific method to solve a problem.

Other scientists experiment for the sake of pure knowlege. Jane Goodall ventured into the wilds of Africa to study the chimpanzee. She sought knowledge and understanding of this creature through observations and experimentation. As a result of her work, much information about the chimpanzee was acquired. More often than not, pure scientific knowledge eventually leads to technological advancement.

Humans have ascended into the vast void of space. Experiments have been designed to investigate the formation of the solar system in an effort to confirm well-established theories. Ongoing experimentation requires scientists to use new knowledge to discredit theories or to confirm scientific laws.

SCIENCE IN HISTORY

Science has played an important role throughout history. The ongoing process of acquiring knowledge through experimentation has solved problems, conquered ignorance, squelched superstitions, and provided inventions to improve the quality of human life. Historians surmise that scientific experimentation has been an essential human activity since the beginning of humankind. Many questions have led the historians to this conclusion.

How did early humans invent and make their tools? How did they learn to hunt, fish, plant seed crops, build dams, make fires, navigate the waters, obtain iron from its ores, make clay pottery, tan leather, domesticate animals, and build wheels? How did they learn to communicate and record events?

A large realm of knowledge was attained prior to the dawn of recorded history. These early discoveries were the crude beginnings of scientific investigation.

Science, as it is understood today, developed during the golden age of ancient Greece. The Greeks learned much from exchanging ideas with the Egyptians, the Babylonians, and others.

Prior to the time of the Greek scientists, supernatural forces were thought to control events in the universe. The Greeks developed and taught the scientific laws that govern the universe. These principles marked an astounding contribution by the Greek culture and have been the foundation for all science to this day.

Scientific investigation keeps everyone at the threshold of future dreams. Your immediate dream is probably a successful experiment in the science lab. The first step to achieving this goal is to learn the secret of scientific success: the scientific method.

SELECTED READING

de Kruif, Paul. *Hunger Fighters*. New York: Harcourt, Brace, 1928.

Eakin, Richard M. *Great Scientists Speak Again*. Berkeley and Los Angeles: University of California Press, 1982.

Gutnik, Martin J. *Electricity: From Faraday to Solar Energy.* New York: Franklin Watts, 1986.

Haber, Louis, *Black Pioneers of Science and Invention.* New York: Harcourt Brace, 1970.

Harre, Rom. *Great Scientific Experiments: Twenty Experiments That Changed Our View of the World.* New York: Oxford University Press, 1983.

Jacobs, Francine. *Breakthrough: The True Story of Penicillin.* New York: Dodd, Mead, 1985.

Lampton, Christopher. *Astronomy: From Copernicus to the Space Age.* New York: Franklin Watts, 1987.

Morris, Richard. *Dismantling the Universe: The Nature of Scientific Discovery.* New York: Simon and Schuster, 1983.

Peters, James A. *Classic Papers in Genetics.* Englewood Cliffs, N.J.: Prentice-Hall, 1959.

Stwertka, Albert. *Physics: From Newton to the Big Bang.* New York: Franklin Watts, 1986.

2 THE RECIPE FOR SUCCESS

How would you describe your style of experimentation in the lab? Are you organized, or are you an unsystematic dabbler?

Scientists have realized that to pass knowledge on to future generations a logical and uniform procedure of inquiry and investigation is necessary. Although inspiration often comes when least expected, and many great scientific discoveries are accidental, most scientific investigation follows a rigorous plan. A systematic approach allows scientists to compare results and duplicate experiments and to attempt to prove the truth or falsity of their beliefs.

This systematic approach, called the *scientific method*, distinguishes scientific inquiry from haphazard, hit-or-miss attempts in the laboratory. Following this method will lead you to a better chance for success in the lab and a feeling of confidence. In addition, once you know how to "do" science properly, you will be more open to those special experiences that make the scientific endeavor exciting: flashes of inspiration and accidental discoveries.

AN EXAMPLE OF THE SCIENTIFIC METHOD

The scientific method is really easy to learn. As a matter of fact, you probably use it quite frequently without even thinking, since it is based on common sense.

Have you ever stumbled into a partially darkened room and flicked the wall switch to find that the lamp does not light? The problem of an unlit lamp has been identified. You study the problem in order to formulate a *hypothesis*—a tentative solution to explain the relationship among observed facts.

You first presume that the lamp cord has come unplugged. Upon observation, you see that it is plugged into the outlet. Next you check to make sure the lamp itself is turned on. It is. You wonder if perhaps you blew a fuse.

To test this hypothesis you try other lamps in the room. They all light up. The fuse was not blown. Finally, you hypothesize that the lamp needs a new bulb. You try a new bulb, and eureka! The lamp works. This sort of thinking and procedure approximate the scientific method.

THE SCIENTIFIC METHOD

IDENTIFY THE PROBLEM

One trait of a successful scientist is an inquisitive mind. A scientist must be able to identify a problem. Well-defined scientific problems can be turned into questions to be answered by systematic experimentation.

For example, how does a green plant use light to produce its food? Why does your heart rate increase when you engage in physical activity? Does the use of drugs by pregnant women lead to birth defects in their children? Does loud noise impair the hearing of listeners? It is very important to define your problem clearly.

There are always problems to be solved in science. In fact, the solutions to these problems often create new problems. Consequently, scientific research is perpetual. Successful experimentation leads to new investigation and to scientific advancement.

STUDY THE PROBLEM

Prior to beginning an investigation, a scientist collects all the information related to the problem. This will save time and effort as he or she is making use of the research already done on a particular subject. It is not necessary to duplicate the work of previous experimentation. You may only need to change the conditions to fit your problem. Libraries and other resource centers are important to scientific research.

FORMULATE A HYPOTHESIS

A hypothesis offers a possible solution to the problem. This proposed solution must be stated very specifically. A scientist knows that this hypothesis is tentative as it may need to be modified or discarded if the evidence does not prove it. A hypothesis, therefore, must be stated so that it lends itself to testing by objective means. The scientist must be able to measure the results.

TEST THE HYPOTHESIS

This plan is called the *experimental design*. A contolled experimental design is often used to test a hypothesis. This method is one in which all conditions are identical except for the single variable to be tested. This variable is called the variable factor or experimental factor.

Suppose you want to determine the effect of water temperature on the respiration rate of goldfish. First, you formulate a hypothesis, or tentative answer to this question. You predict that as the water temperature increases, the respiration rate of the fish will also increase.

Next you set up two tanks of goldfish. One is the experimental tank, the other the control tank. The tanks are identical except for the variable factor, the water temperature. Both tanks contain two goldfish weighing approximately 10 grams each. Two green plants, a gravel floor, and a filter system are included in each aquarium. The fish in both groups are fed 0.1 gram of fish food once each day.

The temperature of the water in the control tank is maintained at 23° Celsius (C). The temperature of the water in the experimental tank is increased by intervals of 5°C at uniform intervals of time. A thermometer is placed in each tank to verify the temperature. The temperature of the experimental tank is increased by adding warm water. Do not allow the temperature of the experimental tank to exceed 60°C. An equal volume of 23°C water is added to the control tank.

To find out the effect of the change in water temperature, you count the number of times the gill covering the goldfish moves in a minute. This gives you the respiration rate of the goldfish. You must record the results for both groups each time new water is added.

You then examine the data collected. Depending on the experimental data, you would conclude that your hypothesis is either correct or incorrect.

OBSERVE AND RECORD DATA

Critical observation is essential to accurately support or contradict the hypothesis. Write down everything you see, noting changes with detailed descriptions.

For instance, the pH paper (acid/base indicator) in your chemistry lab turns a lime-green color when dipped in one solution and a dull blue-green when submerged in another substance. It is not sufficient to write down just "green" for either test. You must record the *exact* colors for each test. Even if the results are not what you expected or seem "wrong," record exactly what you observe. Sometimes the unplanned-for results in an experiment yield even more-significant findings and may establish a hypothesis for a future experiment.

For example, Alexander Fleming discovered penicillin due to results not planned for in an experiment. While experimenting with bacteria, his cultures became contaminated with mold. Instead of discarding the cultures he observed them. He noticed that bacteria were not growing around the mold. This unexpected observation led to the discovery that that particular mold produces a substance, penicillin, that inhibits the growth of bacteria.

In addition to observations, every aspect of the experiment must be recorded accurately. You may use drawings, charts, graphs, tables, notes, calculations, or other visual representations to record your data.

INTERPRET RESULTS AND DRAW CONCLUSIONS

Scientific data mean nothing until they are interpreted and explained. Valid conclusions must be drawn from the experimental evidence or observed facts. A fact is something about which there is no doubt. Facts are based on observations. Data are facts. Conclusions based on facts demonstrated in the experiment may eventually become theories. If the experimental data continue to prove the hypothesis over a period of time, the hypothesis then becomes a *law*.

A confirmed hypothesis is a law. A law explains facts: it is a fact that the more force I apply to an object, the faster it accelerates. A *theory* explains why laws work, why they are the way they are.

From a theory one should be able to derive new results. A theory makes predictions. When verified—or if verified—these predictions verify (or disprove) the theory. Parts of Darwin's theory of evolution are now being revised or challenged because some scientists believe evolution occurs in jumps, not gradually. Still, evolution itself is a fact, so scientists are simply looking for a better theory. A theory can never be conclusively proved, but we may find out something that *disproves* a theory.

You probably will not establish laws in your experimentation, but you can make the most of your facts. It is important to distinguish between observations and interpretations. For example, when observing a beaker of clear liquid, don't assume it is water. Further testing is necessary to establish this fact. Good interpretations are based on accurate experimentation and data collection, not on assumptions.

Now that you understand the scientific method, you are ready to learn more about what it takes to become a master scientist, or perhaps simply a student who no longer feels anxious in the lab, who does well, and who has fun.

SELECTED READING

Beller, Joel. *So You Want to Do a Science Project!* New York: Arco, 1982.

Beveridge, William I. *The Art of Scientific Investigation.* New York: Random House, 1960.

Moorman, Thomas. *How to Make Your Science Project Scientific.* New York: Atheneum, 1978.

Science Fairs and Projects. Washington, D.C.: National Science Teachers Association, 1985.

Smith, Norman. *How Fast Do Your Oysters Grow?* New York: Julian Messner, 1982.

Tocci, Salvatore. *How to Do a Science Project.* New York: Franklin Watts, 1986.

Van Deman, Barry A., and McDonald, Ed. *Nuts and Bolts: A Matter of Fact Guide to Science Fair Projects.* Hammond Heights, Ill.: Science Man Press, 1980.

3
HOW TO SUCCEED ON LAB DAY

Success in the lab does not happen by chance. Success is the result of careful planning before lab day and proper execution of the plan on lab day. Your plan will take into account the amount of time you have, practicing unfamiliar techniques, organizing and recording observations, and developing good lab skills. This plan will allow you to work efficiently, safely, and accurately on lab day. If you have experienced frustration with lab work in the past, a well-developed plan will help reduce or eliminate that frustration. You *can* experience lab success.

FOLLOW DIRECTIONS EXACTLY

The main reason students have trouble in the science lab is that they do not follow directions. Following written directions is a simple three-step process. You must (1) read the directions, (2) understand what you've read, and (3) follow the instructions.

Start by reading the directions through once. Then try to focus on each detail and form a mental image of what the instructions are directing you to do. If you do not understand, reread that section or ask your teacher to explain.

Next, divide the directions into individual steps or processes. For instance, the directions for the experiment may include (1) procedures for setting up the experiment, (2) methods for completing the experimental process, and (3) instructions for recording and interpreting data.

Read and do each step in the first section. Check to see that you have followed the directions exactly; then proceed to the next section. You might find it helpful to check off each step as it is completed.

ORGANIZE YOUR TIME EFFICIENTLY

Read through the directions for the entire experiment *before* you begin. Familiarize yourself with all the material and procedures that you will be using and following during the experiment. Consider the time requirements for each procedure and budget your time. Your lab plan should include an estimate of the time required to complete each section. You may be able to complete two sections of the experiment at one time or work on another section while you are waiting for your first section to be completed.

PRACTICE YOUR PROCEDURES CAREFULLY

Be sure you know how to operate all of the equipment you will be using during your experiment. If you have not used a particular piece of equipment for a long time, practice with the piece of equipment before doing the experiment. Practice any new procedures if you are unfamiliar with the techniques used. The practice will help eliminate errors caused by poor technique and produce more-accurate results.

RECORD YOUR OBSERVATIONS ACCURATELY

Observe and record events that happen during your experiment. Observations are made either directly with your senses or using instruments such as telescopes and balances. Sometimes a computer can be used to record observations, but usually you will record observations with pen or pencil and paper.

Observations of such events as a color change in chemicals, a smell, an eclipse of the moon, or a strange growth in a batch of growth medium form the data for your experiment. Record everything you do and observe, even if it doesn't seem important at the time. Many scientific discoveries were made by chance or are based on unexpected or apparently unimportant observations.

It is important that these data be recorded accurately. A data record sheet (Figure 1) is used for this purpose. Prepare two data record sheets, if not already provided, *before* you begin the experiment. Use one to record your results on lab day and one to submit with your final lab report.

Set aside some space on your data sheet to record data that you are not instructed to record. This additional data might be observed during the lab and could help explain unexpected results.

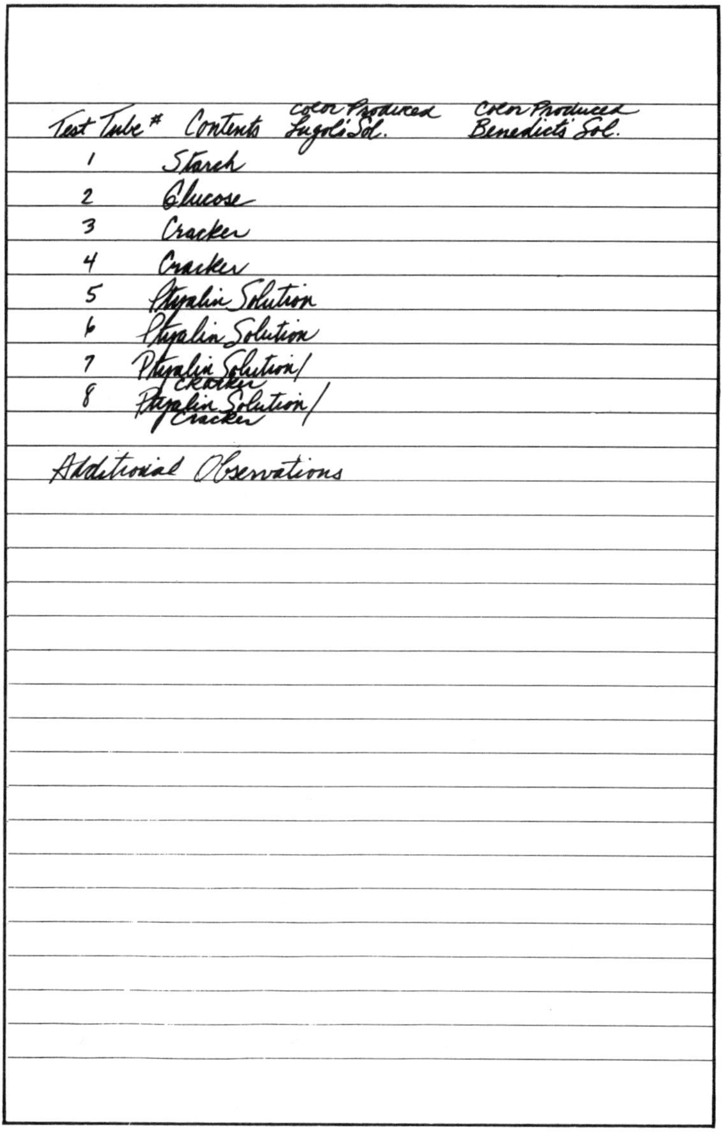

Test Tube #	Contents	Color Produced Lugol's Sol.	Color Produced Benedict's Sol.
1	Starch		
2	Glucose		
3	Cracker		
4	Cracker		
5	Ptyalin Solution		
6	Ptyalin Solution		
7	Ptyalin Solution / Cracker		
8	Ptyalin Solution / Cracker		

Additional Observations

Figure 1. A sample data record sheet

The most common data record sheet is in the form of a chart or table. The area at the bottom of the chart could be used to record additional observations.

At no time should you change the data on your data sheet. If your data don't agree with your expected results, look for reasons that might cause the disagreement. Was your hypothesis correct? Did you follow the directions exactly? Was your equipment functioning properly? You may want to repeat the experiment, a common practice in scientific laboratories.

When you make an original drawing from a microscope observation or a dissection, do not redo the drawing later on. Also, make the drawing immediately. It is hard to remember details after you leave the lab. Hand in your original drawing. You may change something in the copying process.

It is very important to be objective when making your observations. We all have an idea of what to expect in an experiment. Don't allow your expectations to influence your observations. Record exactly what you see, when you see it. Don't assume anything or anticipate your observations.

You will obtain a majority of your lab data from measurements. You will measure length, volume, mass, temperature, or time. For these measurements use the International Standard of Measurement, or the metric system. This system is based on units of 10, so changing units is easy. For example, 125 centimeters is 1.25 meters (100 centimeters are in a meter). Review the metric system and do some practice problems if you feel uncomfortable working in metric.

Measurements should always include units. When possible, take all measurements using the same units. This will make computations easier.

The accuracy of your measurements will directly affect your results and conclusions. The limitations of

CONVERSION TABLE FOR UNITS

Length
1 millimeter = 0.1 centimeter = 0.03937 inch
1 centimeter = 10 millimeters = 0.3937 inch
1 meter = 100 centimeters = 39.37 inches
1 kilometer = 1000 meters = 3280.8 feet
          = 0.621 mile

Mass
1 milligram = 0.0000353 ounce
1 gram = 1000 milligrams = 0.0353 ounce
1 kilogram = 1000 grams = 2.2 pounds

Capacity
1 milliliter = 0.0338 fluid ounce
1 liter = 1000 milliliters = 1.0567 quarts

TABLE OF METRIC ABBREVIATIONS

Length
mm   millimeter
cm   centimeter
m   meter
km   kilometer

Weight
mg   milligram
g   gram
kg   kilogram

Capacity
ml   milliliter
liter liter

your equipment and your skill at using the equipment can cause measurements to be inaccurate. You probably can do nothing about the accuracy of your equipment. You are stuck with what your school has or what you can afford at home. But, you *can* increase your accuracy by becoming more skilled at using your equipment and by being very careful.

Make all measurements at least three times. If the measurements are similar and seem reasonable, find the average.

For example, the width of a block measures 5.2 centimeters, 5.3 centimeters, and 5.2 centimeters in three separate measurements. The average is 5.23 centimeters. This is rounded off to 5.2 centimeters because you cannot claim to be more accurate than your original measurements. If one measurement is completely out of line, drop it from your calculations and average the remaining two. For example, the width of a block measures 5.2 centimeters, 5.3 centimeters, and 5.8 centimeters. Drop the 5.8 centimeters and average the remaining two. The average is rounded off to 5.3 centimeters. If you do this, be sure to explain what you have done. Do not delete the dropped number from the record you are keeping of such measurements.

PRACTICE LAB SAFETY

Lab safety should be one of your main concerns. Are there any dangerous chemicals, equipment, or procedures in your experiment? Remember that accidents do not just happen; they are caused.

The time you spend learning lab safety could prevent a serious lab accident. Most lab work is relatively safe. Procedures have been tried many times. Cautions are included in most updated lab manuals to warn of any possible dangerous materials or procedures. Read and follow these cautions carefully.

Science can be exciting and safe. Always follow these basic rules in the lab:

1. Do all experiments under the supervision of your instructor.

2. Keep your lab area neat and clear of unnecessary material. See Figures 2 and 3.

3. Be familiar with all lab procedures and equipment.

4. Be professional in the lab—horseplay has no place during the lab period.

5. Report all accidents or injuries to your teacher immediately.

6. Clean up any spills—leave your lab area clean. Exercise special precautions when cleaning up caustic materials. See your teacher.

7. Familiarize yourself with the location of safety and fire equipment in the lab—know the classes of fires and how to extinguish them properly.

8. Always wear safety goggles when working with caustic chemicals or when heating liquids. Do not wear contact lenses when working with chemicals.

9. Do not handle gas valves or other equipment that you are not using in your experiment.

10. Wear appropriate clothing—loose-fitting clothing or dangling accessories can be dangerous. Wear a lab coat to protect your clothes. See Figure 4.

Figure 2. Does your work area look like this . . .

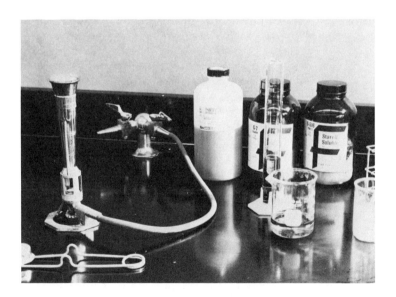

Figure 3. . . . . or like this?

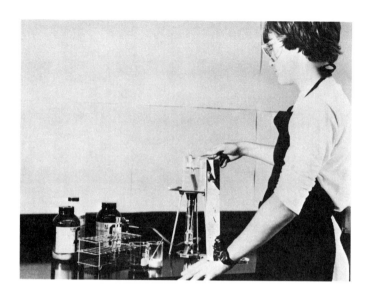

Figure 4. Good work habits are essential for safety and successful experimentation. This student has a neat work area, wears a lab apron and safety glasses, and aims the test tube away, in case the liquid shoots out.

11. Secure long hair.
12. Leave food and beverages outside the lab—food may become contaminated.
13. Dispose of broken glass and lab waste material properly.
14. Use a taste or smell test only when instructed to do so in your lab instructions.
15. Read each label twice—you might read it wrong the first time.
16. Never point a test tube at yourself or another student when heating—the contents may splash out of the tube.

17. Mix chemicals only when directed to do so in the lab procedures.
18. Before inserting glass through a rubber stopper, protect your hands and lubricate the glass. Care must be taken.
19. Dispose of chemicals as directed.

Additional safety information is available through your state or school district's safety manual. Ask your teacher if you can see his or her copy.

DEVELOP PROPER WORK HABITS

Preparing for lab the day before the lab period will save you time in the lab. Your results will be more accurate and the chances of an accident reduced. The plan you develop at home can be executed on lab day. You will be organized and know what to expect.

On lab day gather all the equipment you will need to complete the sections you have scheduled for that day. Practice any unfamiliar techniques. Mix any solutions. Follow all lab directions exactly as they are written.

Keep your lab area clear of unnecessary material. Store your books clear of your lab area. You should have only your lab manual and data record sheet in your lab area.

Before you begin experimenting, make sure your lab equipment is set up properly. Is your equipment clean? Once you complete a procedure or section, remove any equipment you will not be using again.

Do not contaminate your stock chemicals or solutions by returning unused quantities to the container. Estimate the amount needed and discard the excess.

As you experiment, record all events that occur as they occur. Make the necessary measurements and observations. Be aware of any change in color, temperature, or state of matter. Are any gas bubbles produced? These observations may be valuable in drawing conclusions.

Stop your experimenting in time to clean up your lab area and plan for the next day. Can material be stored overnight? Has all the equipment been cleaned and put in its proper place? Did you wipe up all spills?

You need a serious attitude on lab day. Concentrate on your procedures and observations. Don't allow other students to distract you or influence your results. If you develop a good, workable lab plan, you will be successful.

THE BIG DAY

Courtney and Brendon are assigned an experiment in which they must test for carbohydrates in various substances and then investigate the effects of salivary amylase (also called ptyalin) on carbohydrates. Salivary amylase is a chemical found in saliva.

MATERIALS

1. Test tube rack
2. 8 test tubes
3. Test tube holder
4. 2 eyedroppers
5. Goggles
6. 10-milliliter graduated cylinder
7. 50-milliliter beaker
8. 1,000-milliliter beaker
9. Wax marking pencil
10. Hot plate
11. Benedict's solution
12. Lugol's solution
13. 1% starch solution
14. 5% glucose solution
15. Salivary amylase
16. Mortar and pestle

PROCEDURE A:
SETTING UP A WATER BATH

1. Use a hot plate and 1,000-milliliter beaker.

2. Add 500 milliliters of water to the beaker, and bring the water to a boil.

PROCEDURE B:
CARBOHYDRATE TEST

1. Label two test tubes "1" and "2" using the wax pencil.

2. Add 5 milliliters of the starch solution to test tube 1.

3. Add 5 milliliters of the glucose to test tube 2.

4. Use one eyedropper to add 3 drops of Lugol's solution to test tube 1. Observe any color produced. Record your results. Use this eyedropper to measure Lugol's solution only.

5. Use the other eyedropper to add 30 drops of Benedict's solution to test tube 2. Heat the test tube in a water bath for 5 minutes. Observe the color produced. Record your results. Keep this eyedropper for Benedict's solution only.

Caution: Wear goggles when working at the water bath.

PROCEDURE C:
DETERMINING WHICH CARBOHYDRATES ARE IN A CRACKER

1. Label two test tubes "3" and "4" using the wax pencil.

2. Crush a cracker using the mortar and pestle.

Add half the cracker to test tube 3 and the other half to test tube 4.

3. Use one eyedropper to add 3 drops of Lugol's solution to test tube 3. Observe the color produced and record your results.

4. Use the other eyedropper to add 30 drops of Benedict's solution to test tube 4. Heat the test tube in a water bath for 5 minutes. Observe the color produced and record your results.

PROCEDURE D:
DETERMINING WHETHER CARBOHYDRATES ARE IN SALIVARY AMYLASE

1. Label two test tubes "5" and "6" using the wax pencil.

2. Add 5 milliliters of salivary amylase to each test tube.

3. Use one eyedropper to add 3 drops of Lugol's solution to test tube 5. Observe the color produced and record your results.

4. Use the other eyedropper to add 30 drops of Benedict's solution to test tube 6. Heat the test tube in a water bath for 5 minutes. Observe the color produced and record your results.

PROCEDURE E:
DETERMINING WHAT HAPPENS TO THE CARBOHYDRATE IN A CRACKER WHEN SALIVARY AMYLASE IS ADDED

1. Label two test tubes "7" and "8" using the wax pencil.

2. Crush a cracker using the mortar and pestle.

Add half the cracker to test tube 7 and half to test tube 8.

3. Add 15 milliliters of salivary amylase to each test tube. Wait 5 minutes.
4. Use one eyedropper to add 3 drops of Lugol's solution to test tube 7. Observe the color produced. Record your results.
5. Use the other eyedropper to add 30 drops of Benedict's solution to test tube 8. Heat the test tube in a water bath for 5 minutes. Observe the color produced. Record your results.

Courtney and Brendon first read the entire lab. They note that the lab is in five parts testing four different substances with Lugol's and Benedict's solutions. The Benedict's solution must be heated 5 minutes in a water bath. Observations will include colors produced by adding the solution to different substances in eight test tubes.

Courtney developed a data record sheet for the lab and made two copies. This sheet will organize the observations and make interpretations of the results easier.

They must now develop their lab plan so that the lab can be finished within one lab period. Brendon suggests they divide the procedure between them to make the most efficient use of the 45-minute period.

Brendon suggests they ready test tubes 2, 4, 6, and 8 first since the Benedict's solution must be heated for 5 minutes. While they are waiting they will have time to add Lugol's solution to test tubes 1, 3, 5, and 7 and observe the color produced.

Courtney estimates it will take 10 minutes to gather the equipment, 5 minutes to set up the water bath, 10 minutes to add the substances to the test tubes, 10

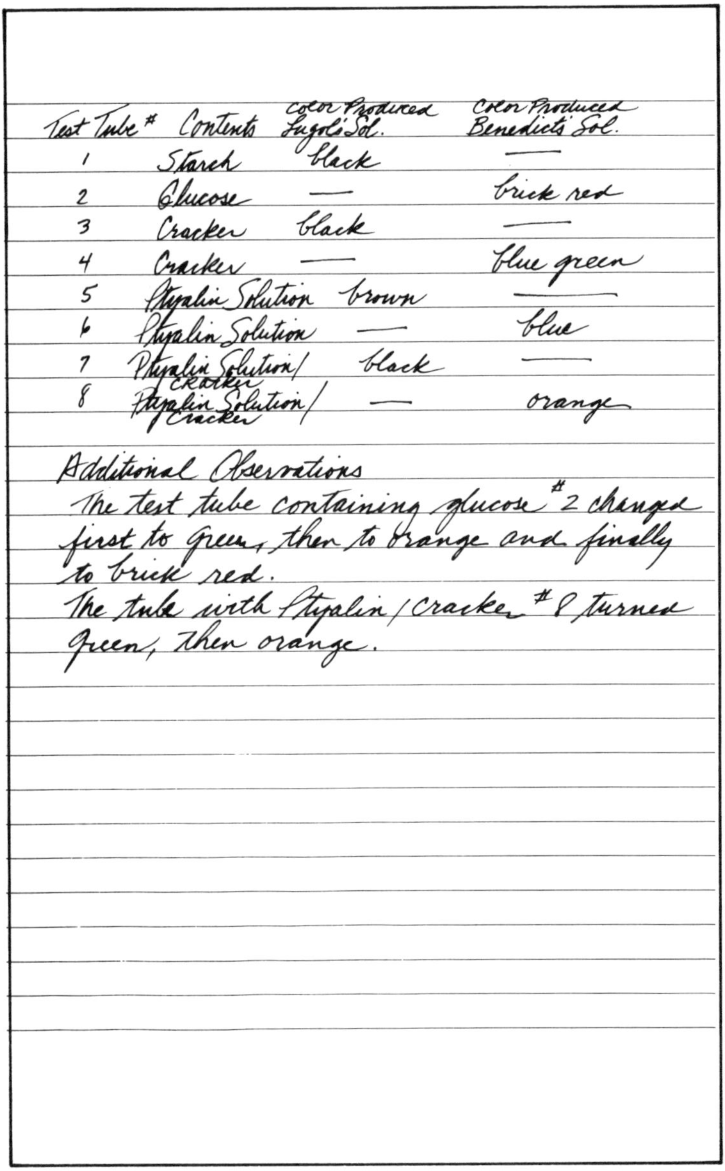

Test Tube #	Contents	Color Produced Lugol's Sol.	Color Produced Benedict's Sol.
1	Starch	black	—
2	Glucose	—	brick red
3	Cracker	black	—
4	Cracker	—	blue green
5	Ptyalin Solution	brown	—
6	Ptyalin Solution	—	blue
7	Ptyalin Solution / Cracker	black	—
8	Ptyalin Solution / Cracker	—	orange

Additional Observations

The test tube containing glucose #2 changed first to green, then to orange and finally to brick red.

The tube with Ptyalin/Cracker #8 turned green, then orange.

Figure 5. The data record sheet filled in with data

minutes to complete the Benedict's test and make all observations, and 10 minutes to clean up the lab area and wash the test tubes and beakers. The lab can be completed within 45 minutes.

As Courtney is setting up the water bath, Brendon will gather all the other equipment, number the test tubes, and measure and add the starch and glucose solution to test tubes 1 and 2. Courtney will crumble a cracker and add it to test tubes 3 and 4. Meanwhile, Brendon will add the salivary amylase solution to test tubes 5 and 6 and crush the cracker to add to test tubes 7 and 8.

Courtney will add Benedict's solution to test tubes 2, 4, 6, and 8 and place them in the water bath. Brendon reminds her to wear her safety goggles and use the test tube holder.

Brendon adds Lugol's solution to test tubes 1, 3, 5, and 7 and observes and records the colors produced.

Courtney finishes the Benedict's test, then observes and records the colors produced. Figure 5 shows the filled-in data record sheet. Brendon starts to clean up but not before Courtney has a chance to observe all the test tubes.

Once the lab is clean and the glassware is washed and put away, they begin to write their conclusions.

Courtney and Brendon approached the lab experiment with a plan. They executed their plan on lab day. They were successful in completing the lab safely and efficiently. They experienced success in the science laboratory. Courtney and Brendon planned to use the next class period to analyze their data and write their conclusions.

SELECTED READING

Gerlovich, Jack A. *Better Science Through Safety.* Ames: Iowa State University Press, 1981.

4
THE TOOLS AND TECHNIQUES OF THE TRADE

Scientific progress is directly related to the development and refinement of scientific equipment. Human beings have been able to walk on the moon, cure disease, develop computers, and communicate around the world because the right equipment has been available.

Your success in the science laboratory depends in part on how well you understand and use scientific equipment. Successful manipulation will reduce human error and produce more-accurate results.

This chapter introduces you to various kinds of lab equipment and outlines the correct procedures for using it.

> **Always wear safety goggles when working in the laboratory.**

TECHNIQUES FOR MIXING AND DISPENSING AND FOR WORKING WITH GLASS

HOW TO WORK WITH GLASSWARE

- **Use only borosilicate glass (Pyrex, Kimax, etc.) if your experiment requires heating a substance in a glass container.** This glassware can be heated with little danger of cracking. Soft-glass apparatus such as a graduated cylinder breaks when heated or filled with hot liquids. See Figure 6.
- **Handle hot glassware carefully.** Use beaker thongs or test tube holders even after the glass is removed from the flame, since glassware remains hot long after heating is stopped. Set

Figure 6. Ordinary jars have no place in the laboratory.

Figure 7. *How to handle hot glassware*

glassware on a hot pad or test tube rack until it has cooled. See Figure 7.
- **When heating a beaker or flask over an open flame, set the container on a wire gauze pad.** Make sure the flame does not lap over the edge of the pad.
- To clean glassware, wash it with soap and water and let it air-dry. **If you use a solvent for cleaning, make sure the flask is completely dry or the solvent may explode. Work in an area with proper ventilation.**
- **Discard or repair cracked or chipped glass.** Firm, repeated stroking with a wire screen or triangular file can smooth sharp edges.
- **Never mix explosive compounds in glass containers.**
- **Always wear goggles when cutting, polishing, or heating glassware.**
- Always wrap glass tubing with a cloth towel and use glycerine if inserting it into a stopper.

HOW TO MIX SOLUTIONS

As you experiment, you will be using different kinds of solutions in varying concentrations. A *solution* is made by dissolving a solid or liquid (*solute*) in a liquid (*solvent*). Sometimes your teacher will premix your solutions; sometimes you will be required to mix your own.

When you mix your solutions, you will be adding either a solid or liquid to another liquid. This will give you a certain percentage of the solute dissolved in the solvent. For example, a 15% solution of salt is prepared by adding 15 grams of salt to enough water to equal 100 milliliters. Always use distilled water.

> **Caution: Wear your safety goggles.**

How to Mix a Percentage Solution Using Liquids
1. Measure the volume of the solute equal to the percentage of the desired solution in a graduated cylinder (Figure 8).
2. Add enough solvent (usually water) to equal 100 milliliters (Figure 9).

Caution: Always add acid slowly to water. Do not add water to acid. It could cause spattering. Wear goggles.

Example: To make a 50% solution of sulfuric acid, first measure 50 milliliters of sulfuric acid. Add the acid slowly to 50 milliliters of water.

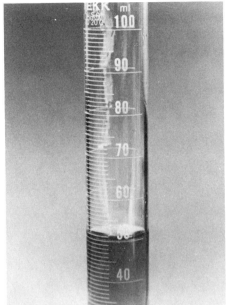

Figure 8. To make a 50% solution using liquids, first measure out 50 ml of solute . . .

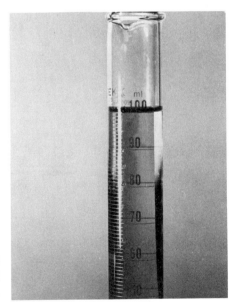

Figure 9. . . . then add enough solvent to equal 100 ml.

*How to Mix a Percentage Solution
Using Solids with Liquids*
1. Weigh in grams the mass of the solute equal to the percentage of the desired solution (Figure 10).
2. Add enough solvent (usually water) to equal 100 milliliters (Figure 11).

Example: To make a 5% glucose solution, weigh out 5 grams of glucose. Add water until the volume equals 100 milliliters.

Figure 10. *(left) To make a 5% solution using solids, first weigh out 5 g of solute . . .*
Figure 11. *. . . . then add enough solvent to equal 100 ml.*

[50]

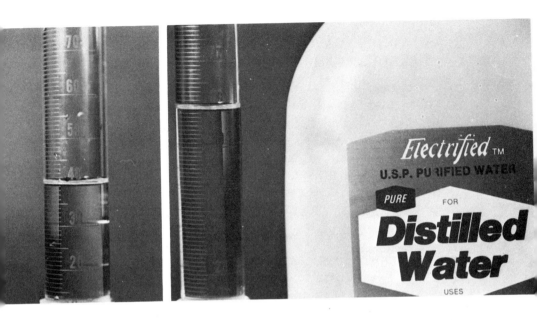

Figure 12. (left) To reduce a 60% solution to a 40% one, measure 40 ml of the original solution . . . **Figure 13.** . . . then add distilled water to bring the volume to 60 ml.

How to Reduce the
Concentration of a Solution
1. Measure in a graduated cylinder a volume of the original solution equal to the concentration of your new solution (Figure 12).

2. Add distilled water to increase the volume to equal the percentage of the original solution (Figure 13).

Example: To reduce a 60% solution to a 40% solution, measure 40 milliliters of the original solution. Add distilled water to bring the volume to 60 milliliters. You now have a 40% solution.

HOW TO POUR
A LIQUID FROM
A REAGENT BOTTLE

A reagent bottle is used to store liquid chemicals and solutions. *Pour* all such liquids from this bottle using a stirring rod. This helps prevent the liquid from contaminating the sides of the receiving container and possibly the pourer. Never *return* a liquid to its stock bottle once it has been removed. Make a good estimate of how much of the chemical you need before you pour it, since you should discard any excess. Make sure you are wearing safety goggles.

1. Remove the glass stopper from the reagent bottle with the back of the index and middle fingers and continue to hold the stopper (Figure 14).

2. Pick up the bottle with the label toward you (Figure 15).

Figure 14. To pour liquid from a reagent bottle, (1) remove the glass stopper...

Figure 15.
. . . (2) Pick up the bottle with the label toward you . . .

Figure 16.
. . . and (3) Pour the liquid along a glass stirring rod.

3. Pour the liquid slowly along a glass stirring rod inserted in the middle of the receiving container (Figure 16).

4. Set the bottle down and replace the glass stopper.

5. Discard any excess liquid not needed for your experiment. Your teacher will tell you how to discard the excess.

HOW TO USE A MORTAR AND PESTLE

The mortar and pestle is used to pulverize (reduce to a powder) chemicals or other substances. Many pulverized chemicals dissolve more readily in liquid, and powdered substances react more easily. See Figure 17.

Figure 17. The mortar and pestle

1. **Make sure you are wearing goggles.**
2. Place the substance in the mortar.
3. Use the pestle to break the large pieces into smaller pieces with an up-and-down action.
4. Pulverize the chemicals by using a circular grinding action.
5. Remove the powder using a spoon or spatula. Do not blow into the mortar to remove the residue. The powder could be blown into your face.
6. Clean the mortar and pestle with soap and water.

HOW TO CUT AND BEND GLASS TUBING AND SOLID GLASS RODS

Glass tubing is used to attach rubber tubing to rubber stoppers in flasks and other apparatus. These connectors can vary in length and can be bent at various angles to accommodate different needs. Solid glass rods are used as stirring rods.

Glass tubing and rods usually come in long lengths. To obtain the length and shape you need, you will have to *cut* the tubing or rod, *bend* it by heating it, and *fire-polish* the cut ends.

How to Cut Glass Tubing or Rods
1. Wear goggles.
2. Place the tubing or rod on a flat surface.
3. Use a triangular file or glass-cutting knife to make a deep scratch in the glass. Do this by moving the file or knife away from you with a single stroke. Excess downward pressure could shatter the glass.
4. Protect your hands from cuts by wrapping a towel or cheesecloth around the tubing or rod on both sides of the scratch (Figure 18).

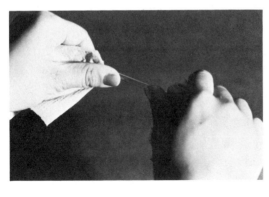

Figure 18. Protect your hands when you break glass tubing.

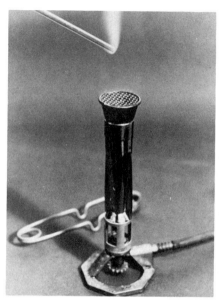

Figure 19.
Fire-polishing
glass tubing

5. Place your hands on each side of the scratch with your thumbs beside the scratch. The scratch should point away from you.

6. Bend the tubing or rod toward you while applying moderate pressure. Push with your thumbs while pulling with your fingers. The tubing or rod should snap at the scratch.

7. The ends of the tubing or rod now need to be smoothed (fire-polished). Heat the ends in the edge of a burner's flame using a rotary motion. Continue heating until the ends become rounded. Heating too long will constrict the tubing. See Figure 19.

8. Glass remains hot for a time after heating. Lay it on a hot pad and be careful not to touch the heated end.

How to Bend Glass Tubing
1. Wear goggles.
2. Cut and fire-polish the tubing.
3. Stopper one end of the tubing.
4. Place a flame spreader on the burner and light the burner.
5. Heat the tubing in the place you want it to bend by rotating it in the flame until it begins to sag or bend. Heat a 4-centimeter area.
6. Remove the tubing from the flame.
7. Place the tubing on a hot pad to make sure it is bent on an even plane.
8. Blow lightly into the unstoppered end of the tubing while gently pulling the two ends toward you as you apply pressure against the pad. The tubing should bend where heated. See Figure 20.
9. Don't touch the heated area of the tubing.

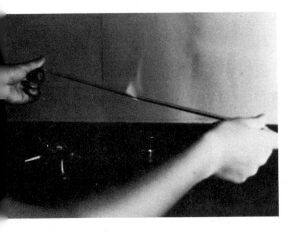

Figure 20.
Heating glass tubing to bend it

HOW TO INSERT GLASS INTO A RUBBER STOPPER

It is sometimes necessary to insert glass tubing, thermometers, or thistle tubes into holed rubber stoppers (see Figure 21). They can then be fitted into flasks and other apparatus. This is relatively simple if you follow the correct procedure.

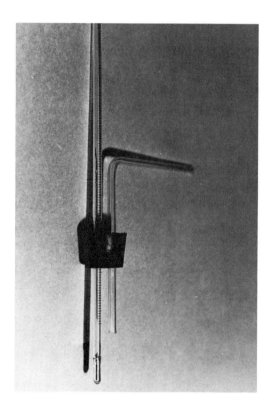

Figure 21. A thermometer and glass tube inserted in a stopper

1. Wear goggles.
2. Select a holed stopper to match the size of the tubing, thermometer, or thistle tube.
3. Lubricate the glass and stopper with water, silicone, grease, or glycerine. This makes it easier to slide the glass into the stopper.
4. Wrap cheesecloth or cloth towels around the glass and the stopper, to lessen the danger of cuts if the glass breaks.
5. Hold the glass as close as possible to the end you plan to insert into the stopper.
6. Keep your hand holding the stopper out of line with the glass. If the glass broke, it could be forced into your hand.
7. Insert the glass into the stopper using a side-to-side rotating motion. Apply moderate force. Excessive force could cause the glass to break. If the glass doesn't insert easily, relubricate the glass and the stopper.
8. Insert the glass to the proper length.
9. If you want to remove the glass from a stopper, apply the lubricant to the glass on the side opposite the direction you plan to remove it.
10. Reverse the inserting action.
11. If you cannot remove the glass easily, cut the stopper with a razor blade or knife, on a flat surface.

TECHNIQUES FOR MEASURING

HOW TO USE A BALANCE

The balance measures mass, and each type is used pretty much the same way. Balances differ mostly in the number of pans and the degree of accuracy. Familiarize yourself with the balance you will be using and locate the various parts *before* you use it (see Figure 22).

1. Place your balance on a level surface. (Some balances have level indicators and leveling screws on the legs.)
2. With all poises on zero, make sure the balance is "in balance." This is when the balance indicator is on zero or swinging equal distances above and below zero. If not "in balance," make adjustments using the knurled balancing nut.

Figure 22. *Two-pan balance (left); beam balance (right)*

3. Place the mass to be measured on the balance pan. For two-pan balances, place the mass on the left pan and add the weights to the right pan. Chemicals and liquids can corrode the pan, so select a piece of paper (folded along the edges to prevent spillage) for chemicals or a container for liquids. These must be weighed empty. This empty-container mass will be subtracted after the chemicals or liquids are added and weighed.

4. On a beam balance use a spatula or spoon to transfer dry chemicals from the bottle to the balance. **Never touch chemicals with your hands.** Move the poises or weights along the beams to restore balance. Slots are provided on the heavier beams. The weights must be in the slot.

5. Add the heaviest weights first, until the arm falls below the zero point. Back off the weights to the next lower weight. The balance arm will now be above the zero point. Continue this procedure with the remaining beams until one beam remains. The last (smallest-weight) beam has a slide poise. Slide it along the beam until balance is obtained.

6. Add the total weights of all the poises used. This is the mass in grams. Remember to subtract the weight of the paper or container if weighing chemicals or liquids.

7. Return the poises to zero and remove the mass.

8. See your teacher if you are still having problems.

HOW TO USE A COMPASS

Use a compass to draw a circle or part of a circle (an arc). See Figure 23.

1. Determine the radius of the circle to be drawn.
2. Open the compass to the desired radius along a metric ruler.
3. Insert the pointed end of the compass and make a circle with the pencil end.

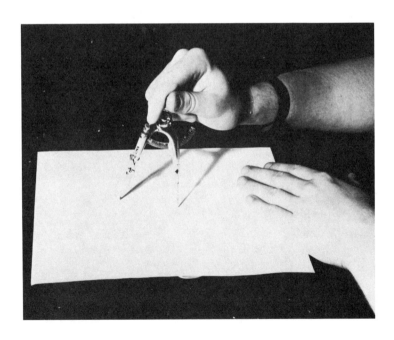

Figure 23. Drawing a circle with a compass

Figure 24. Measuring angles with a protractor

HOW TO USE A PROTRACTOR

A protractor is used to measure angles in degrees. See Figure 24.

1. Determine the angle to be measured.
2. Place the circle with the crosshairs at the vertex of the angle (point where two sides meet).
3. Read the angle in degrees using the correct scale. The 0 should be on the side the angle opens (right or left).
4. Extend the lines of the angle if they are too short.

HOW TO USE A METRIC RULER

Always use a metric ruler in the lab. See Figure 25. Metric rulers vary in size from 15 to 30 centimeters. Use a meterstick for larger measurements.

1. Familiarize yourself with the markings on your ruler. The larger-numbered markings divide the ruler into centimeters. Each centimeter is divided into 10 millimeters, which are not numbered. The fifth millimeter is longer than the other markings, so it is easy to find the midpoint of a centimeter.

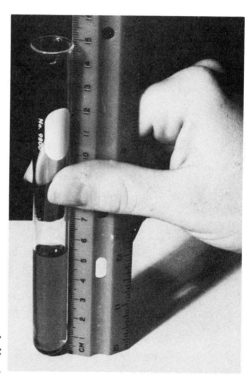

Figure 25. *Always use metric units in the lab.*

2. Start your measurement 1 centimeter from the end of the ruler (edges may be worn with use).
3. Align your eye at the opposite side of the ruler from the object being measured.
4. Count the total number of centimeters.
5. Count the number of millimeters from the last centimeter to the end of the object.
6. Subtract 1 centimeter (remember, you started measuring 1 centimeter from the end).
7. Record the number of centimeters and millimeters (for example, 5.2 centimeters).
8. Repeat the procedures to check for accuracy.

HOW TO USE GRADUATED CYLINDERS

A graduated cylinder is used to measure volumes of liquids in milliliters. Most graduated cylinders are made of glass which produces a curved surface at the surface of the liquid called a *meniscus* (Figure 26). A plastic ring around the cylinder will help prevent breaking if the cylinder is knocked over.

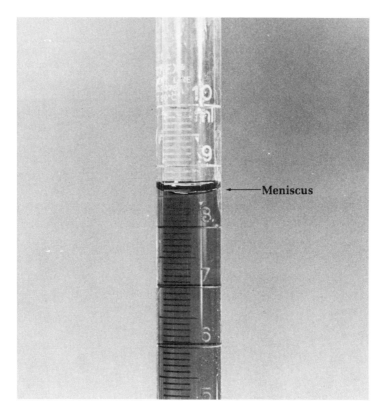

Figure 26. *The volume of a liquid in a graduated cylinder is read at the meniscus.*

> **Caution: Wear your safety goggles.**

1. Determine the value of each division on your cylinder. Two scales may be present: one starting with small numbers at the bottom for measuring liquids, one starting with small numbers at the top for pouring specific amounts of a liquid.
   a. Locate two consecutively numbered lines.
   b. Subtract to find the interval between the lines.
   c. Count the number of lines between the consecutive lines and add one.
   d. Divide the number of lines (step c) into the interval (step b). This is the value of each line in milliliters.

2. Pour the liquid to be measured, into the cylinder.

3. Set the cylinder on a level surface.

4. With your eye level to the surface of the liquid, read the bottom of the meniscus.

5. Read the volume to the nearest line.

HOW TO USE A THERMOMETER

A thermometer measures temperature. Scientists use the Celsius scale (also called centigrade). Water boils at 100°C and freezes at 0°C. See Figure 27.

1. Select the thermometer which has a large enough degree range for your experiment. Most of the time 110°C to −20°C will do.
2. Familiarize yourself with the scale on your thermometer. Most thermometers are divided by 1°C lines.
3. Place the thermometer in the center of the liquid being measured. Use a clamp and ring stand if the liquid is to be heated (see Figure 28). Do not allow the thermometer to touch the side or bottom of the container.

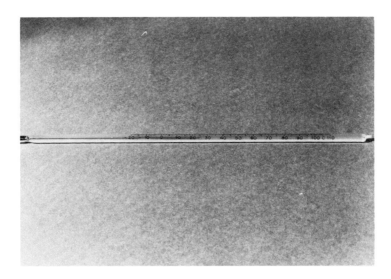

Figure 27. *A metric thermometer*

Figure 28.
How to measure the temperature of a liquid

4. When the column of mercury or alcohol stops rising, read the temperature to the nearest ½ degree.

5. Record the measurement.

6. **Caution: Thermometers usually contain either alcohol or mercury. If you break a mercury thermometer, *don't pick up* the mercury with your hands. It is poisonous and can be absorbed through the skin. Use a piece of paper to collect the mercury into one drop and place it in a glass container. Ask your teacher how to dispose of it.**

TECHNIQUES FOR HEATING SUBSTANCES

HOW TO USE A LABORATORY BURNER

A lab burner, or Bunsen burner, is used as a heat source for a variety of lab activities. By adjusting the gas and air flow, an inner blue cone is produced. The hottest part of the flame is in the outline of the blue cone. See Figure 29.

1. Make sure the main gas valve is off. This will prevent dangerous buildup of gas before you light the burner.

2. Attach your burner to the main valve using a flexible rubber hose. Recheck your attachment points. Also check your hose for leaks.

3. Place your burner on a level lab table back from the edge. Secure the burner so it won't tip over. Familiarize yourself with the way to adjust the gas flow. Some burners use the main valve; others have adjustments on their base.

Using a Match

4. Light the match. **Caution: Do not hold your face over the burner.**

5. Turn on the gas.

6. Bring the match up to the top of the burner's barrel to ignite the gas.

Using a Spark Lighter

4. Turn on the gas.

5. Place the lighter over the burner.

6. Strike the lighter to ignite the gas.

Figure 29. *Heating glass over a bunsen burner*

Figure 30. *Adjusting the flame of a bunsen burner*

7. Adjust the air intake and gas flow until a blue cone appears. Open the air intake to allow more oxygen if the flame is yellow. Close the air intake to reduce oxygen if the flame is blue and no cone is visible. If the flame keeps going out, reduce the gas flow. See Figure 30.

8. Adjust the height of the flame by using the gas flow adjustment. The flame should not lap over the edge of the equipment being heated.

9. Turn off the main gas valve to extinguish the flame.

10. **Caution: Keep loose clothing and hair away from flame. Never leave your burner unattended. Always check to see that the main gas valve is off when you leave your lab station.**

HOW TO HEAT SUBSTANCES IN A TEST TUBE

Test tubes are used for heating a variety of materials. This can be done using direct heat or using a water bath. A water bath allows the material (including flammable substances) to be heated evenly. Do not use an open flame with flammable substances.

> **Wear safety goggles.**

1. Set up the burner.
2. Set up the test tube(s), test tube holder(s), and test tube rack. Be sure the test tubes are of the heat-resistant type.
3. Light the burner and adjust the flame.
4. Pick up the test tube with the holder.
5. **Point the tube opening away from you but not toward anyone else. Its contents could splash or shoot out of the tube.** See Figures 31 and 32.
6. Heat the tube by moving it back and forth through the flame just above the blue cone. Heating in one place could cause the contents to shoot out as the contents are heated too quickly.
7. Heat the tube the required time or until the reaction is complete.
8. Set the hot tube in the test tube rack to cool (Figure 33).
9. Turn off the burner if you are through with it.

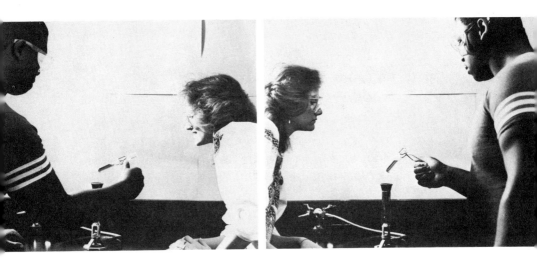

Figure 31. (left) The WRONG way to heat liquid in a test tube. **Figure 32.** The right way to heat liquid in a test tube—pointing the opening of the tube AWAY from you and your lab partner.

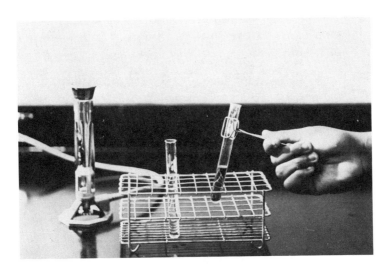

Figure 33. Cooling hot test tubes

TECHNIQUES FOR COLLECTING AND SEPARATING SUBSTANCES

HOW TO USE AN EVAPORATING DISH

An evaporating dish is used to separate solids dissolved in a liquid and to evaporate water contained in a solid. Solutions placed in the dish evaporate, leaving the dissolved solids in the dish.

1. Set up the apparatus shown in Figure 34 using a ring stand, iron ring, wire gauze pad, burner, and (heat-resistant) evaporating dish.

2. Wear goggles.

3. Light the burner. Adjust the flame height so it doesn't lap around the sides of the wire gauze pad.

4. Heat the solution until it has all evaporated. The solid will remain in the dish. **Heat all solvents except water under a hood or in a well-ventilated area.**

5. Turn off the burner and allow the dish to cool.

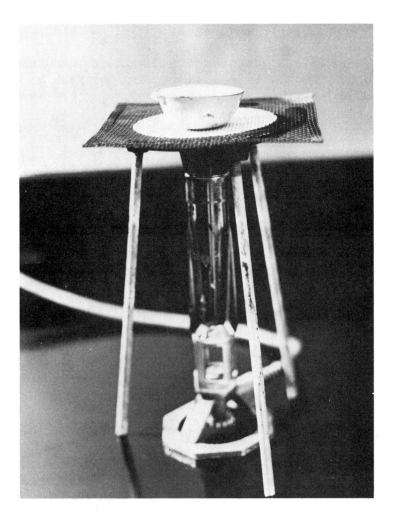

Figure 34. Heating solids in an evaporating dish

HOW TO USE A FUNNEL AND FILTER PAPER

A funnel and filter paper are used to separate substances suspended in liquids. A *suspension* is a mixture containing undissolved particles of a substance. When liquids are poured through the paper, the *filtrate* (liquid part) passes through while the solid parts remain on the paper as a residue.

1. Make sure you are wearing safety goggles.
2. Select the proper-size filter paper for the funnel being used.
3. Fold the paper in half twice. Tear off a corner on the open end. Open the paper and form a cone. See Figure 35.
4. Place the paper in the funnel.
5. Set up the apparatus shown in Figure 36 using a ring stand, iron ring, beaker, and the funnel with paper.

Figure 35. Inserting a filter-paper cone into a funnel

 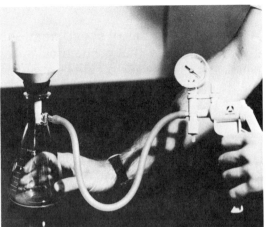

Figure 36. (left) Apparatus for filtering liquids
Figure 37. A hand-operated vacuum pump aspirator can be used to decrease filtration time.

6. Seal the paper by pouring some solvent through it and pressing it gently against the filter wall using a stirring rod. This prevents air from getting in and slowing the filter process.

7. Pour the liquid to be filtered into the funnel using a stirring rod. Be careful not to punch a hole in the paper.

8. The filtrate will drip into the beaker. Touch the funnel stem to the beaker side to prevent splashing. If the liquid is not retained in the funnel stem, the stem is dirty or the paper is not sealed properly.

9. You can decrease the filtering time using suction. An aspirator or other device can be used (see Figure 37).

HOW TO COLLECT A GAS

When you are doing an experiment, the production of a gas is one sign that a chemical reaction has taken place. An ideal way to collect a gas is by measuring the displacement of water by the gas. Refer to Figure 38 for steps 1-4; to Figure 39 for steps 5-8.

> **Since some gases are explosive, be very careful. Wear goggles.**

1. Fill a pneumatic trough with water until it covers the tray insert.
2. Fill a gas bottle with water and cover the mouth with a glass plate.
3. Carefully invert the bottle and lower it into the trough. Be careful not to allow any water to escape. If air bubbles occur in the bottle, repeat steps 2 and 3.
4. Once the bottle's mouth is under the water in the trough, remove the glass plate.
5. Position the bottle over a hole in the tray.
6. Place the rubber tubing from the gas-generating apparatus into the hole below the bottle.
7. The gas generated will force the water out of the bottle and replace it.
8. Stopper the bottle while it is still inverted. The gas can now be analyzed.
9. Always discard the first bottle collected since it contains mostly air from your gas-producing apparatus.

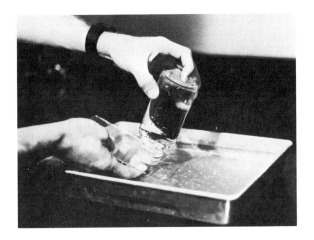

Figure 38. To collect a gas, first insert a water-filled bottle into a pneumatic trough . . .

Figure 39. . . . then connect the gas-generating apparatus to the bottle. Gas will force out the water from the bottle, leaving you with a bottle full of gas.

HOW TO IDENTIFY OXYGEN, HYDROGEN, AND CARBON DIOXIDE

Many chemical reactions produce oxygen, hydrogen, or carbon dioxide as by-products. In some experiments these are exactly what you are trying to produce. How, then, can you detect their presence? A simple test can be conducted.

1. **Wear goggles.**
2. Collect any suspected gas in a container (using the methods described in the previous section).
3. Keep the container inverted or stoppered until you are ready for the test.
4. Light a wood splint.
5. To test for hydrogen or carbon dioxide, insert the burning splint into the mouth of the container. If the splint goes out, the gas is probably carbon dioxide. If the splint causes a popping noise (and flame) the gas is hydrogen (see Figure 40).

    Caution: Hydrogen is an *explosive* gas. It can cause an explosion, or flames may shoot into or out of its container during this test. Do not place your hands or face in line with the mouth of the container. Do not collect *large* quantities of this gas. Do not use a container with thin walls.

6. To test for oxygen, blow out the splint. Insert the glowing splint into the container. If it bursts into flames, the gas is oxygen (see Figgure 41). Keep your hands and face away from the mouth of the bottle.

Figure 40. Testing for hydrogen or carbon dioxide using a burning splint. *Be very careful when doing this test. Keep your hands and face away from the mouth of the bottle. Wear protective goggles.*

Figure 41. Testing for oxygen using a blown-out splint. *The same warning applies as for the previous test.*

TECHNIQUES FOR OBSERVING

HOW TO DISSECT PLANTS AND ANIMALS

Dissection is a useful method of exploring and viewing the structure of plants and animals. A variety of dissecting tools are used for different types of dissections (see Figure 42).

Although all dissections are different, the following hints will be useful when you dissect:

1. Dissect only in a dissecting tray.

2. Never cut or remove any structure unless directed to remove it.

3. **Be careful not to get preservative in your eyes. If somehow you do, wash your eyes with cold water at once and tell your teacher.**

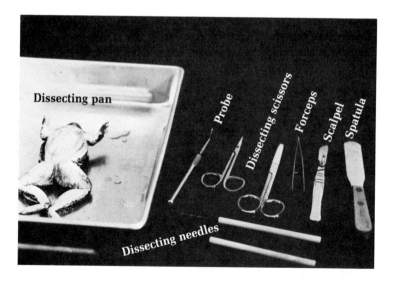

Figure 42. *Dissection tools and frog*

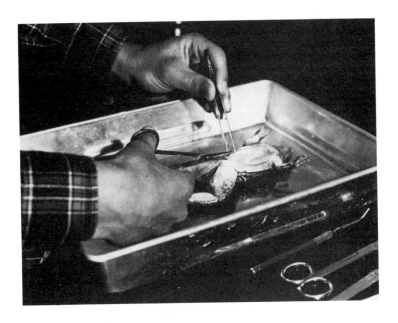

Figure 43. *The correct way to use dissecting scissors.*

4. Use upward pressure when cutting with scissors. Keep the bottom blade as parallel as possible to the line of your cut. Don't cut too deep. See Figure 43.

5. Use a dissecting needle to loosen and separate tissue.

6. To keep material overnight, keep it moist or place it in a plastic bag or refrigerator.

7. Wear goggles.

Selected Reading
Berman, William. *How to Dissect: Special Projects for Advanced Study.* 4th edition. New York: Arco, 1984.

HOW TO USE A MICROSCOPE

The microscope is used to magnify very small objects so that you can see them. A whole new world of living things becomes visible when you use a microscope.

Microscopes vary in magnification and operation. You should become familiar with the scope you are using before you begin using it (see Figure 44). For practice, you may want to prepare a slide using the letter e from a newspaper. Figures 45 and 46 compare magnifications of a specimen.

1. Place the object to be viewed on a slide, add a drop of water, and cover it with a cover slip.
2. Raise the body tube by turning the coarse adjustment knob toward you.
3. Place the slide on the microscope stage and position the object to be viewed over the stage opening.
4. Secure the slide by using the stage clips.
5. Place the lowest-power objective in the viewing position.
6. Looking from the side, lower the body tube until it is *just* above the slide.
7. Look into the eyepiece. Bring the object into focus by raising the tube.
8. Bring the object into sharp focus using the fine-adjustment knob.
9. By moving the slide, center the object in the field of view.
10. Looking from the side, put the high-power objective in the viewing position.
11. Bring the object into sharp focus using the fine-adjustment.

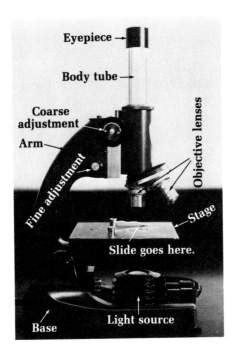

Figure 44. The parts of a microscope

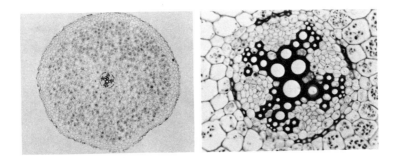

Figure 45. (left) Ranunculus root under the microscope (originally magnified ten times but enlarged in this picture). **Figure 46.** The same view as in Figure 46 but under ten times the magnification (originally magnified one hundred times).

Helpful Hints

1. Clean your microscope lens only with lens paper.

2. The image seen in a microscope is reversed and inverted. Move the slide in the opposite direction from which you want the image to move.

3. Never turn the body tube toward the stage while looking into the eyepiece. You may hit the slide with the objective.

4. Only transparent objects can be viewed using a compound light microscope (see Figure 47).

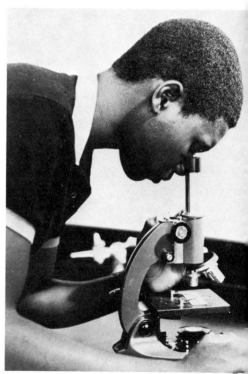

Figure 47. *Only transparent objects can be viewed using a compound light microscope.*

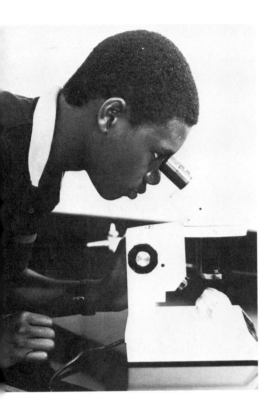

Figure 48. Use a stereo dissecting microscope to view opaque objects.

5. Use a binocular microscope to view objects that are not transparent (see Figure 48).

Selected Reading
Curry, Alan; Grayson, Robin; and Hosey, Geoffrey. *Under the Microscope.* New York: Van Nostrand Reinhold, 1982.

Headstrom, Richard. *Adventures with a Microscope.* New York: Dover, 1941.

Johnson, Gaylord; Bleifeld, Maurice; and Beller, Joel. *Hunting with the Microscope.* New York: Arco, 1980.

HOW TO USE A TELESCOPE

The telescope is used to view objects at a distance. It is most commonly used to view the moon, planets, or stars.

Telescopes are either reflectors or refractors (see Figures 49 and 50). Reflectors use mirrors to do most of the magnifying; refractors use lenses.

1. Place the telescope's tripod on a level surface.
2. Spread the tripod legs so that the scope is steady.
3. Look through the finderscope and locate the object to be viewed.
4. Center the object in the finderscope.
5. Focus your telescope by holding the tube and pushing or pulling the focusing tube until the object is in focus.
6. **Caution: Never look directly at the sun with a telescope.** Figure 51 shows a safe setup for viewing the sun.

Selected Reading
Brown, Sam. *All About Telescopes.* Cambridge, Mass.: Sky Publishing, 1976.

King, Henry C. *The History of the Telescope.* New York: Dover, 1979.

Thompson, Allyn J. *Making Your Own Telescope.* Cambridge, Mass.: Sky Publishing, 1980.

Worvill, Roy. *Stars and Telescopes for the Beginner.* New York: Taplinger, 1980.

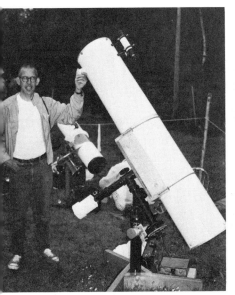

Figure 49. (left) Ten-inch (25-cm) reflecting telescope built by amateur astronomer George Cogswell, who is holding the eyepiece.
Figure 50. Five-inch (13-cm) refracting telescope built by Tal Mentall (wearing the hat).

Figure 51. The safest way to observe the sun is to project the image from the eyepiece of a telescope onto a white board, as is being done here. The man at the left is holding the main telescope.

HOW TO WORK WITH A PETRI DISH

A Petri dish is used to grow bacteria, yeast, mold, or fungi (see Figure 52). These microorganisms are grown on a nutrient medium (agar) in an ideal environment. This environment allows for maximum growth, making it easy to study these organisms. For the best results, follow sterile (organism-free) techniques.

1. Sterilize your Petri dish by boiling it in water for five minutes or placing it in an *autoclave* (if available). An autoclave allows you to heat substances to high temperatures without boiling. It is used to sterilize material.

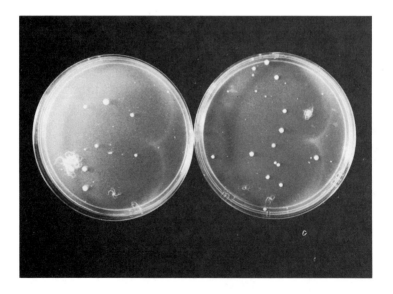

Figure 52. *Petri dishes containing bacterial growths*

2. Prepare the nutrient agar, autoclave if available, and add it to the Petri dish.
3. Cover the dish immediately. It is now ready to use.
4. The dish can be exposed to microorganisms in a variety of ways. You can simply open the dish and expose it to the air. You can touch an object to the surface of the agar, or you can transfer known organisms using a wire loop.
5. Grow the exposed organisms at room temperature placing the dish upside down. This will prevent water that condenses on the lid from dripping on your organisms.
6. **Caution: Do not open the dish if you don't know which culture is growing. Your dish could contain a harmful bacteria. This could be *very* dangerous.**

5
R2D2 HAS NOTHING ON YOU

"Now what do I do with all of these numbers that I recorded during the science experiment?"

Learning how to calculate, interpret, and present your data is essential to laboratory success. A knowledge of basic math skills will help you to complete this task.

SIGNIFICANT FIGURES AND ROUNDING OFF

Let's begin by looking at the idea of *significant figures*. These are the digits known to be accurate. The data in an experiment are only as precise as the least precise data. Therefore, the derived quantity in an experiment can have only as many significant figures as the least precise quantity of the data in the calculation.

A chemist weighing three substances obtains weights of 1.21 grams, 3.196 grams, and 2.16 grams.

What is the total weight? Simple addition yields 6.566 grams. But two of the three weights are accurate to three significant figures, while one is accurate to four significant figures.

Since the calculated quantity can have only as many significant figures as the least accurate quantity from the data used in the calculation, the sum would be rounded off to three significant figures. The usual practice for rounding off is to round the number up if the next digit is 5 or greater, *down* if the next digit is less than 5.

The chemist would record the weight as 6.57 grams, which is 6.566 rounded off to three significant figures.

SCIENTIFIC NOTATION

Scientists frequently have to use very large and very small numbers. The planet Pluto is 3,660,000,000 miles from the sun; Avogadro's number, the number of atoms in a gram of an element, is about 60,000,000,000,000,000,000,000; a hemoglobin molecule is about 0.00000007 centimeter in diameter; the earth is about 5,000,000,000 years old. The number of zeros makes reading and writing these numbers in the above forms difficult, so *scientific notation* is used.

Each number is expressed as a small number between 1 and 10, followed by another, very large number expressed in powers of 10. A hemoglobin molecule is 6.7×10^{-8} centimeters in diameter. The number 3,660,000,000 is 3.66 multiplied by a number containing nine zeros: 3,660,000,000 = 3.66 times 1,000,000,000—in powers of 10, 1,000,000,000 = 10^9. Scientific notation expresses the distance to Pluto as 3.66×10^9 miles.

Numbers can be multiplied by multiplying the

base numbers and adding the exponents. For example:

$$(2 \times 10^5) \times (4 \times 10^3) = 8 \times 10^8$$
$$(3 \times 10^{-3}) \times (2 \times 10^{-6}) = 6 \times 10^{-9}$$
$$(5 \times 10^{-7}) \times (7 \times 10^8) = 35 \times 10^1 = 3.5 \times 10^2$$

In division the base numbers are divided and the exponents subtracted. For example:

$$\frac{6 \times 10^4}{2 \times 10^2} = 3 \times 10^2$$

$$\frac{8 \times 10^5}{2 \times 10^7} = 4 \times 10^{-2}$$

$$\frac{3.6 \times 10^{-7}}{6 \times 10^{-5}} = 0.6 \times 10^{-2} = 6 \times 10^{-3}$$

To square these numbers (multiply them by themselves), you square the base numbers and multiply the exponents by 2. For example:

the square of $3 \times 10^7 = 9 \times 10^{14}$

the square of $4 \times 10^{-6} = 16 \times 10^{-12} = 1.6 \times 10^{-11}$

You take the square root of the base number and divide the exponent by 2 to derive the square root of the number. If the exponent is not a multiple of 2 (an even number), shift the decimal point so that the exponent is divisible by 2. For example:

$$\sqrt{9 \times 10^{12}} = 3 \times 10^6$$

$$\sqrt{4.9 \times 10^{-5}} = 7 \times 10^{-3}$$

(remember, $4.9 \times 10^{-5} = 49 \times 10^{-6}$)

Scientific notation makes cumbersome calculations, with very large and very small numbers, much more manageable.

PERCENTAGES

During scientific experimentation it is often necessary to calculate percentages. For example, chemistry labs often require you to determine the percentage of a substance in a solution. Genetic labs often use percentages to indicate the occurrence of a given trait in a population.

Percent means "hundredths." Percent can be thought of as a fraction. The numeral preceding the % symbol represents the numerator, and the % symbol indicates a denominator of 100. For example,

$$50\% = 50/100$$

Percent can also easily be related to decimal numerals. Since the % symbol means hundredths, it indicates two decimal places. To write a percent as a decimal numeral, remove the % symbol and move the decimal point two places to the left. For example,

$$50\% = 0.50 = 0.5$$

Scientists usually record data using decimal numerals. Fractions and mixed numbers should be converted to decimals to make calculations easier. However, percents are used when comparing data. For example, do we need a water solution of 10% salt or 40% salt?

USING SCIENTIFIC EQUATIONS AND FORMULAS

An equation is used to calculate the value of an unknown number, represented by a letter (a symbol, sometimes called a variable). The equation is formed by translating two equal facts into algebraic expressions. At least one of these facts will contain the unknown number. One expression is written equal to the other. The equation can then be solved.

For example, suppose you have to find the average speed at which you ride your bike. We can use a formula to calculate this unknown value:

$$\frac{\text{Distance}}{\text{Time}} = \text{Velocity}$$

Your watch tells you that you traveled for 2 hours. Your odometer indicates you traveled 30 miles.

$$\frac{30 \text{ miles}}{2 \text{ hours}} = 15 \text{ miles/hour}$$

Your average cycling speed is 15 miles/hour.

TABLES

A table is an excellent format for your data. Tables are easy to construct and facilitate analysis of your data.

For example, suppose you wanted to determine the effect of the "home team advantage" on the basketball shooting average of individual team members. You will need to keep careful records of the goals scored by each team member in all games—home and away. You would compute the average goals per home game and per away game for each player. The information could be collected in a table such as Table 1.

TABLE 1

PLAYER	GAMES PLAYED		TOTAL HOME GOALS	TOTAL AWAY GOALS	AVERAGE HOME GOALS*	AVERAGE AWAY GOALS*
	Home	Away				
Dave	7	6	109	90	8	7
Chuck	7	7	129	89	9	6
Steve	6	7	107	84	8	6
Jason	7	7	139	93	9	7
Phil	7	7	123	93	9	7

*Averages are rounded off to the nearest goal.

Table 1 is easy to read. It allows you to compare the number of goals scored at away games with the number of goals scored at home games for each player. It also allows you to compare individual players in terms of goals scored. From these data you can conclude, for example, that playing on the home court is an advantage for all five of the players.

When constructing a table, include all data necessary for the experiment. Clearly label each column and row, and neatly organize the figures. A computer can be used to construct tables.

GRAPHS

Graphs are another valuable way to present data. A *line graph* is used to represent relationships between variables. A *bar graph* is useful for comparing the size of quantities in statistics. A *circle* graph is used to represent the relation of the parts to the whole and to each other.

A line graph has a horizontal (x) axis and a vertical (y) axis. These axes are perpendicular number lines that intersect at the (0,0) point (see Figure 53). To locate any point for our graph, we need a pair of numbers that can be plotted on coordinate axes.

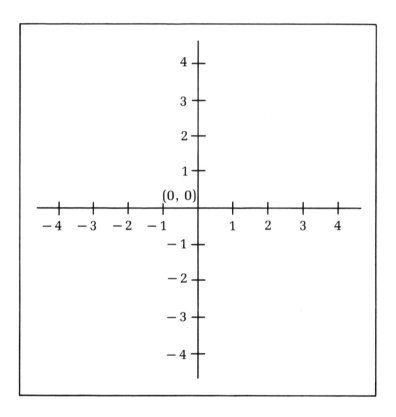

Figure 53. *A line graph*

To plot a point on the graph, we use two intersecting perpendicular lines, one from the x-axis, the other from the y-axis. The x-coordinate is the number associated with the point where the perpendicular line intersects the x-axis. The y-coordinate is the number associated with the point where the perpendicular line intersects the y-axis. These x- and y-coordinates are expressed as an ordered pair of numbers in this manner: (3,5). The x-coordinate is written first. These two numbers are plotted in Figure 54.

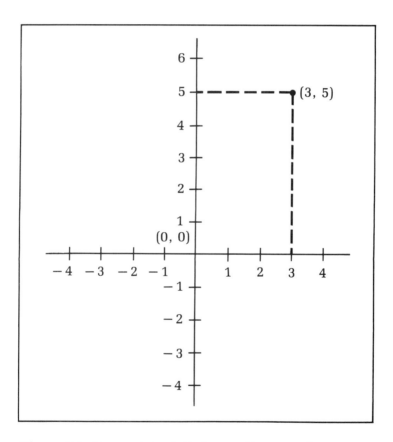

Figure 54. *One point plotted on a line graph*

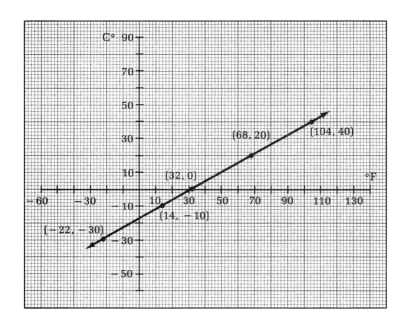

Figure 55. *The relationship between Fahrenheit and Celsius temperature readings plotted on a line graph*

Figure 55 shows the relationship between Fahrenheit and Celsius temperature readings. The formula for the conversion of Celsius to Fahrenheit temperature readings is: F = 1.8C + 32

As a first step to constructing this graph, we must develop a table of values showing the relationship between the variables in the formula:

F	C
32°	0°
68°	20°
104°	40°
14°	−10°
−22°	−30°

We then must select a convenient scale (range) for variables on each axis. We place and label this scale along each axis. The values in the table are the pairs of coordinates that we plot on our graph. We draw a line to connect these points. This line is the graph of the formula. The formula $F = 1.8C + 32$ is the title for our graph.

To construct a bar graph we draw an x-axis and a y-axis on graph paper. Again, select a scale for the values being compared and label the scale. An example of a bar graph is shown in Figure 56. This graph indicates New York City's average temperature in "degrees Fahrenheit" for each month of the year.

To construct a circle graph, first construct a table that includes the given facts and the fraction or percent of the whole which each quantity comprises. Then calculate the number of degrees representing each fractional part or percent by multiplying 360° by each fraction or percent:

%	Calculation	In Circle
25%	$\dfrac{25 \times 360}{100}$	90°
50%	$\dfrac{50 \times 360}{100}$	180°
12%	$\dfrac{12 \times 360}{100}$	43°
13%	$\dfrac{13 \times 360}{100}$	47°

Figure 56. *This bar graph shows the average temperature in each month of the year.*

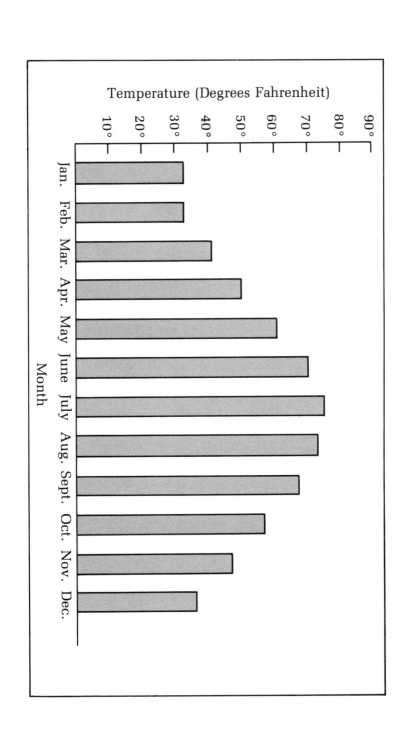

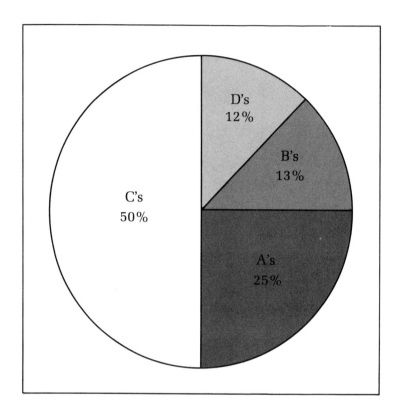

Figure 57. A circle graph, also known as a pie chart

Figure 57 shows the distribution of grades students attained on a science achievement test.

Most computers have the ability to display graphics, and many programs exist that enable you to tabulate data and convert those data into graphs of all types.

In summary, the skill that separates the technician from the scientist is the ability to analyze data. This skill involves examining the information collected during the experiment and drawing conclusions based on that information. As explained in this chapter, the techniques for calculating, interpreting, and presenting your data will help you to be a success in the lab.

6 YOU'RE READY TO DO A SCIENCE PROJECT

Each year, hundreds of thousands of students all over the world complete science projects. Although science projects are often required in science classes, much can be gained by doing the assignment besides fulfilling the requirement. This holds true for successful science students as well as for those who are nervous about science, who dislike science, or who plan never to study science again.

Probably the best reason for doing a science project is the pride and self-satisfaction you will feel after completing a well-designed project. Your systematic approach using the scientific method may have enabled you to discover something not known before. Who knows? Maybe your research will eventually lead to an important discovery.

If you are interested in science but don't do well on tests, you may be able to demonstrate your science knowledge and skills in a science project using creative thinking and other skills such as library research and artistic talent, to enhance your project (see Figure 58). Besides being personally rewarding, it can improve your grade!

Doing a science project can teach you skills that may be useful in college or in a job. For example, the fact that you took on this extra task will indicate your responsibility and dedication to your work. Both are characteristics of a good student and employee.

If you enter your project in a science fair, you may win a prize, of course. Winners may receive college scholarships or grants, cash prizes, or trips to scientific installations in government or industry. You will also receive public recognition for your award-winning achievements. Your school will probably honor you in some way.

"BUT WHAT SHOULD I DO?"

Sometimes teachers assign topics for projects, but often you may select your own topic. If you have no idea what you want to do, start by deciding which general area of science interests you. Do you prefer biology, chemistry, physics, geology, or astronomy? You may be spending considerable time on your project, so choose an area that really interests you and that you will enjoy. After doing this, you will have to narrow the topic to something manageable. For instance, you can't expect to crack the DNA code in your science project.

Figure 58. *John Vartanian, a contestant at a science fair held on Long Island, New York*

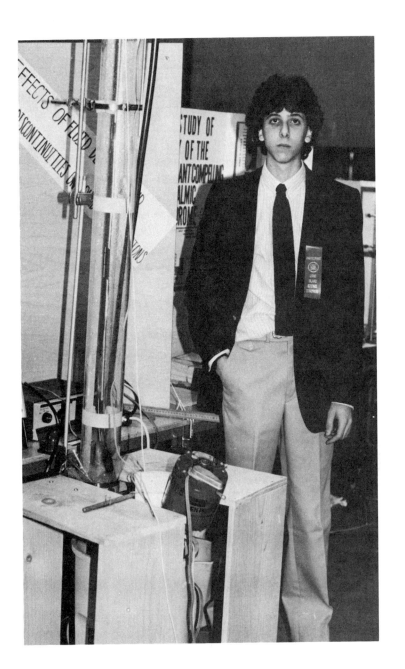

In order for your experiment to be significant, you must follow the scientific method, outlined in Chapter 2. The initial steps in this method require you to define and study a problem and then formulate a hypothesis—a possible explanation for some phenomenon.

You can begin a project with either a question or a statement, which then can be answered or proved or disproved. For instance, "What happens when you smoke a cigarette?" is unsatisfactory because many answers are possible. There are social implications, physical effects, psychological effects, and so forth. A better question would be, "Does smoking one pack of cigarettes per day increase your chances of getting lung cancer?" This question should lead to a definite answer. From this question you can formulate a hypothesis. You might say, "Smoking a pack of cigarettes a day causes cancer." The hypothesis must also be *manageable*. The above hypothesis might take thirty to forty years to prove or disprove!

A hypothesis must be able to be tested by objective means. You must be able to measure the results. The smoking experiment is possible because you can test the hypothesis by setting up an experiment with control and experimental groups. You could determine the number of cases of lung cancer in a population of people who smoke a pack of cigarettes, and you could find out the number of cases of lung cancer in a control group of nonsmokers. You could measure the difference between incidences of lung cancer in the two populations.

LET'S LOOK IT UP!

Let's assume you have an idea for your project or have decided upon a general field of science to explore.

Your next step should be taken in the library. No matter what area of science you have selected, you can be almost certain that some research has already been done in that area. The work, procedures, and results of other investigators have been recorded and are probably housed in either your school library or a public library. Reading this scientific literature will help you to conduct your own investigation. Therefore, learning how to locate materials in the library is an essential skill.

Begin with the card catalog, because it lists every book in the library. The books are categorized three ways, so there are three cards for each book. One card is filed alphabetically by the author's last name. The second card is filed alphabetically according to the title of the book. The third card is filed alphabetically by subject.

It is important that you look at the date of publication for each book. This information is noted on the card following the name of the publishing company. This date will give you a general idea of the relevancy of the material in the book. You may have to read classic studies conducted in the past. For current findings, however, remember that science moves at a fast pace. For instance, a book about recombinant-DNA techniques published two years ago may now be out of date due to recent discoveries.

Excellent sources for current information are journal and magazine articles. The *Reader's Guide to Periodical Literature* lists approximately 155 magazines and journals. Volumes of the *Reader's Guide* are located in the reference section of the library. Each volume covers articles written during a specified period of time. A hardcover edition is published annually, and a softcover edition comes out every two months.

Articles are entered under the author's name and under as many subjects as each article requires. Each entry gives all the bibliographic information necessary for locating the article. Each entry contains many abbreviations. There is usually a key to the abbreviations used.

Both the card catalog and the *Reader's Guide* cross-reference information. That means sometimes when you are looking up a particular subject you will see no article or book titles listed but instead find instructions to look under another heading or other headings.

There are other directories in addition to the *Reader's Guide*. The *Science Citation Index* deals only with scientific topics. An excellent index of scientific journals is *Ayers Dictionary of Publications*. If you are interested in scientific investigation worldwide you may want to consult *Ulrich's International Periodical Directory*.

Reading and understanding the articles in some of the specialized scientific periodicals will probably be difficult, perhaps even impossible, so you may need to read special journals called abstracts that feature short reviews of articles that appear in other journals. The abstracts help you to understand some of the complicated articles.

You will want to take notes while you sift through books and articles, so you should establish an organized method for taking notes. The most standard system is to take notes on 3 × 5 index cards. The advantage of using cards is that you can arrange them easily. Some students use a notebook to record this information. You may want to use a computer, if you have one. The index card should contain the title of the book or article, the name of the author, the name of the publisher or publication, the date of publication, and of course your notes. From information on the card you should be able to prepare your bibliography.

TYPES OF PROJECTS

The way you display your science project is very important. Your research and findings should be carefully developed and presented in an eye-catching, interesting manner.

An enjoyable and easy science project is an artistically arranged pictorial presentation. You can use your imagination to display photos, sketches, and diagrams to explain scientific theories, to relate the history of a scientific device or innovation or to teach a successful new medical procedure. For example, you may do an experiment to determine what effect the absence of phosphate would have on the growth of a plant. You could take a series of pictures to compare a plant given phosphate with one that is not given phosphate. These pictures could be used to illustrate and support your conclusions.

Magazines, newspapers, college and university literature, and pamphlets and booklets printed by industries or government agencies are valuable sources for pictures. For instance, suppose your project addresses alternative forms of energy, such as geothermal and solar energy. You may be able to get pictures related to this subject from a local gas or electric company or from an educational institution that does research in this area.

The essential ingredients for a winning pictorial display are large, colorful pictures arranged in an interesting and logical manner. You will need clever captions to draw attention to each picture as well as a concise summary that explains the ideas conveyed. Tables, charts, and graphs are also valuable tools for displaying scientific findings.

Models are an effective way to express your science project (see Figure 59). (Examples include model jet engines, replicas of prehistoric organisms,

Figure 59. Scientists and engineers make models; you can too.

skeletons hearts, lungs, and other parts of the human body.) Other students use clay, cardboard, plastic, papier-mâché, wood, metal, plaster of paris, styrofoam, paints, and other materials to create original models. Original models are usually preferred as they require much more ingenuity, effort, and hard work to build. A working model better illustrates the scientific principles involved in your science project. A solar cell that uses light energy from a light bulb and converts it into electrical energy to run an electric motor would be more useful in explaining solar power than a stationary model. A model should have a title and labels, as well as a written explanation of the scientific process illustrated by the model.

The scientific survey provides another impressive format for your science project. Surveys are also useful for determining if there is a relationship between certain factors. You may wish to compare seemingly unrelated sets of data. For instance, you could survey your classmates to determine how many take vitamins daily. These data could then be correlated to the number of absences for each student during a designated period of time. You might conclude that those who don't take vitamins have significantly more absences than those who do. Of course, sometimes the results of the survey are not what you predicted. This is also valuable information.

It is not always necessary for you to collect the data yourself. Sometimes it may be advantageous to use statistics compiled by others. A variety of sources are available for statistical information. The ones you use depend upon the nature of the survey.

For instance, suppose you wanted to determine if the energy shortage and the awareness of limited resources have influenced consumers' choice of cars. You could go to all of the car dealers in your town. You would need to record the numbers of each type of car sold in the past, let's say, ten years. By graphing the data, you could draw conclusions about trends during the past decade. You might also be able to predict what particular car or type of car will be the most popular, if present trends continue.

The psychological survey usually addresses behavior. Data on animal behavior are usually collected by experimentation. By administering a questionnaire, you may obtain data related to human behavior. The questionnaire is given to a representative sample of the population being studied. The responses are analyzed, and conclusions are drawn from these data.

For example, suppose you want to find out who

has a better self-image, ninth-grade boys or ninth-grade girls. You would devise a questionnaire to answer this question. Then you would administer it to a significant number of subjects—ninth-grade boys and girls. The greater the number of subjects, the more reliable the results will be. This is true for all scientific research. The number of boys and girls sampled should be equal.

This questionnaire could be simple. You might write twenty to thirty statements. The instructions could ask the student to respond yes to each statement which describes him/her, and no to each statement that does not, and "undecided" if the student doesn't know. Sample statements could include:

1. I am well liked by my classmates.
2. I am an interesting person.
3. Most people think I am stupid.

THE WRITTEN REPORT

After finishing your research and drawing your conclusions, you will have to write a report. The report explains your methods. Some students experience anxiety just thinking about writing this report. This is probably because they have no plan. If you have a logical and practical plan for the written report, then following it to completion generally goes quite smoothly.

Here is what the report should contain:

1. *Title and Cover Page.* Your title should convey the scientific concepts your project addresses. It should reflect what is in your report. The cover page should include the title, your name,

and other required information for identification.

2. *Table of Contents.* This follows the title page. It lists everything contained in the report, including pictures, tables, graphs, and other visual data.

3. *Procedures and Methods.* Explain how you did your project. Organize this section around the steps of the scientific method to describe each component of your scientific investigation. Begin with a clearly stated hypothesis and successfully complete each step to the results and conclusions.

4. *Conclusions.* The summation of your work should always relate directly back to the original problem or hypothesis. Often your scientific investigation will support your hypothesis. Sometimes the hypothesis is proven incorrect. The conclusions do not always suggest a solution to the problem. Your findings may pose questions for future investigation. Always give a detailed interpretation of your conclusions.

5. *Bibliography.* Include a list of the books and articles you used. For books, identify the author, the name of the book, the location (city) and name of the publisher, and the year of publication. For magazines, include the author's name, the title of the article, and the name, volume, and number of the magazine. List the entries alphabetically by the author's last name. Books are listed like this:

> Black, Jason. *The Evolution of Man.* New York: Smith Publishing Co., 1982.

Woods, Caroline. *Recombinant DNA Techniques.* Boston: Worldwide Publishing Co., 1981.

Articles in periodicals are usually listed like this:

Wallace, Amy. "How to Use Classical Conditioning with Your Dog." *Journal of Animal Care.* Vol. 6. June 1986.

SELECTED READING

Apfel, Necia H. *Astronomy and Planetology: Projects for Young Scientists.* New York: Franklin Watts, 1983.

Beeler, Nelson F., and Brankley, Franklin M. *Experiments in Optical Illusion.* New York: Thomas Y. Crowell, 1951.

Beller, Joel. *Experimenting with Plants.* New York: Arco, 1985.

Brown, Vinson. *Building Your Own Nature Museum for Study and Pleasure.* New York: Arco, 1985.

Byers, T.J. *20 Selected Solar Projects.* Englewood Cliffs, N.J.: Prentice-Hall, 1984.

Clark, John G., and Stone, Harris. *Science Project Puzzlers.* Englewood Cliffs, N.J.: Prentice-Hall, 1981.

Cobb, Vicki. *Chemically Active! Experiments You Can Do at Home.* New York: Lippincott, 1985.

Cobb, Vicki. *Magic . . . Naturally! Science Entertainments and Amusements.* New York: Lippincott, 1976.

Cobb, Vicki. *Science Experiments You Can Eat.* New York: Harper and Row, 1982.

Dunbar, Robert E. *The Heart and Circulatory System: Projects for Young Scientists.* New York: Franklin Watts, 1984.

Gardner, Martin. *Entertaining Science Experiments with Everyday Objects.* New York: Dover, 1981.

Gardner, Robert. *Energy Projects for Young Scientists.* New York: Franklin Watts, 1987.

Gardner, Robert. *Ideas for Science Projects.* New York: Franklin Watts, 1986.

Gardner, Robert. *Kitchen Chemistry.* New York: Julian Messner, 1982.

Goodwin, Peter H. *Engineering Projects for Young Scientists.* New York: Franklin Watts, 1987.

Goodwin, Peter. *Physics with Computers.* New York: Arco, 1985.

Gutnik, Martin A. *Ecology: Projects for Young Scientists.* New York: Franklin Watts, 1984.

Gutnik, Martin A. *Genetics: Projects for Young Scientists.* New York: Franklin Watts, 1985.

MacFarlane, Ruth B. *Collecting and Preserving Plants for Science and Pleasure.* New York: Arco, 1985.

McKay, David W., and Smith, Bruce G. *Space Science Projects for Young Scientists.* New York: Franklin Watts, 1986.

Pawlicki, T. *How to Build a Flying Saucer and Other Proposals in Speculative Engineering.* Englewood Cliffs, N.J.: Prentice-Hall, 1980.

Provenzo, Eugene F., Jr., and Provenzo, Asterie Baker. *Rediscovering Photography.* La Jolla, Calif.: Oak Tree, 1980.

Research Problems in Biology: Investigations for Students (3 volumes). Biological Sciences Curriculum Study. New York: Oxford University Press, 1976.

Schmidt, Victor E., and Rockcastle, Verne N. *Teaching Science with Every Day Things.* 2d ed. New York: McGraw-Hill, 1982.

Schulman, Elayne; Craigo, Ken; Griffiths, William F.; and Megna, Denise. *Science Projects with Computers.* New York: Arco, 1985.

Stone, George K. *Science Projects You Can Do.* Englewood Cliffs, N.J.: Prentice-Hall, 1963.

Stong, C.L. *Scientific American Books of Projects for the Amateur Scientist.* New York: Simon and Schuster, 1960.

Tocci, Salvatore. *Chemistry Around You: Experiments and Projects with Everyday Products.* New York: Arco, 1985.

White, Laurance B., Jr., and Broekel, Ray. *Optical Illusions.* New York: Franklin Watts, 1986.

Wood, Elizabeth A. *Science from Your Airplane Window.* New York: Dover, 1975.

See also the titles listed at the end of Chapter 2.

PERIODICALS

To keep yourself informed on developments in science and technology, you may want to read the following publications:

Discover
High Technology
Natural History
Popular Science
Science
Science News
Scientific American
Smithsonian
Technology Review

GLOSSARY

autoclave—laboratory equipment used to sterilize different materials, making them free of living bacteria.

bar graph—a graph useful for comparing quantities.

circle graph—a graph useful for expressing the relation of the parts to the whole. Also called a pie chart.

experimental plan—the best method for testing variable factors or hypotheses.

filtrate—material that has passed through a filter.

hypothesis—a statement that suggests an explanation for a relationship between observed facts.

law—a confirmed hypothesis; it explains facts.

line graph—a graph used to explain relationships between variables.

meniscus—the curved surface of a liquid column.

percent—hundredths; one part of a hundred.

scientific method—the specific procedure that scientists use to solve problems.

scientific notation—a form that scientists use for writing very large or very small numbers.

significant figures—in a quantity, the number of figures that are accurate.

solute—the substance dissolved in a solvent.

solution—a homogeneous liquid with variable composition consisting of a solute and a solvent.

solvent—material that dissolves a solute.

suspension—a heterogeneous mixture that contains both a liquid and solid.

theory—an explanation of a law or laws that also makes predictions.

INDEX

Italicized page numbers indicate illustrations.

Aspirators, 77
Autoclaves, 90–91, 123

Bacteria, observation of, 90–91
Balances, 60–61
Bibliographies, 112, 117–18
Bunsen burners, 70–71

Carbohydrates, test for, 38–41, *42*, 43
Carbon dioxide, 80, *81*
Collecting and separating techniques
    evaporation, 74, *75*
    filtering, 76–77
    for gases, 78, 79, 80, *81*
Compasses, 62
Conclusions, 23–24, 117

Data calculation, 93–97
Data interpretation, 23–24
Data presentation
    graphs, 98–105
    tables, 97–98
Data recording, 23, 29, *30*, 31, 42
Direction-following, 27–28
Dissecting microscopes, 87
Dissection, 82–83
Drawings from observation, 31

Electric motors, 14
Equations, 97
Evaporating dishes, 74, 75
Evolution, theory of, 24
Experimental plans and designs, 21–22, 123

Faraday, Michael, 14
Filtering liquids, 76–77
Filtrates, 76, 123
Fleming, Alexander, 23
Formulas, 97

Gases
  collection of, 78, *79*
  identification of, 80, *81*
Glass tubing and rods, 55–59
Glassware, use of, 46–47
Goodall, Jane, 15
Graduated cylinders, 66–67
Graphs, 98–102, *103*, 104, 123

Heating of substances, 70–72, *73*
Henry, Joseph, 14
Hydrogen, 80, *81*
Hypotheses, 14–15, 24, 123
  formulation, of 20, 21, 22
  testing of, 21–22, 110

Immunization, 13, 15

Lab burners, 70–71
Lab procedures
  and direction-following, 27–28
  example of, 38–41, *42*, 43
  observations, recording of, 29, *30*, 31
  practice of, 28–29
  safety and, 33–34, *35*, 36–37
  time, efficient organization of, 28
  work habits, 37–38
Laws, 24, 123
Library sources, 111–12
Liquids, pouring of, 52–53

Measurements, accuracy of, 31, 33
Measuring instruments
  balances, 60–61
  compasses, 62
  graduate cylinders, 66–67
  metric rulers, 64–65
  protractors, 63
  thermometers, 68–69
Mendel, Gregor, 14–15
Meniscus, 66, 67, 123
Metric rulers, 64–65
Metric system, 31, *32*
Metric thermometers, 68

[126]

Microscopes, 84, *85*, 86–87
Models, 113–14
Mortar and pestle, 54

Note-taking, 112

Observation, 23, 29, 31
Observation techniques
    dissection, 82–83
    microscope, 84, *85*, 86–87
    petri dish, 90–91
    telescope, 88, *89*
Oxygen, 80, *81*

Pasteur, Louis, 13, 15
Penicillin, 23
Percentage solutions, 48, *49*, 50–51
Percentages, 96, 123
Pneumatic trough, 78, *79*
Problems, 20–21
Protractors, 63

Reagent bottles, 52–53
Rubber stoppers, 58–59
Rulers, metric, 64–65

Safety, 33–34, *35*, 36–37
Salivary amylase, 40–41
Science projects, 107–8, *109*
    choice of, 108, 110
    reference work in, 110–12

types of, 113–16
written reports, 116–18
Scientific experiments
    in history, 15–16
    importance of, 13–14
    reasons for, 14–15
Scientific method, 19–24, 110, 123
Scientific notation, 94–96, 123
Separating substances. *See* Collecting and separating techniques
Significant figures, 93–94, 124
Solutes, 48, 124
Solutions, 48–51, 124
Solvents, 48, 124
Sun, viewing of, 88, *89*
Surveys, 115–16
Suspensions, 76, 124

Tables, 97–98
Telescopes, 88, *89*
Test tubes, heating substances in, 72, *73*
Theories, 24, 124
Thermometers, 58–59, 68–69
Time, use of, 28

Variable factors, 22
Vartanian, John, *109*

Water baths, 39

ABOUT THE AUTHORS

Diane Wallace has been a high school reading specialist in Delaware since 1979. She also teaches at the University of Delaware. Diane has presented numerous workshops on teaching reading in science, received many awards for her achievements, and written textbooks. She lives with her husband and her young daughter in the house they built in Hockessin, Delaware.

For fifteen years, Philip L. Hershey has taught biology at William Penn High School, where he also has developed curricula for numerous courses. He was selected by the National Association of Biology Teachers, Biology Teacher of the Year for Delaware in 1986. Phil also coaches various sports, and in 1981 was selected track coach of the year in Delaware. He lives with his wife and three children in New Castle, Delaware.

Diane Wallace and Philip L. Hershey have also authored the teacher resource book for *Biology: Living System*, a textbook.